THE PLANT LIFE
OF
SNOWDONIA

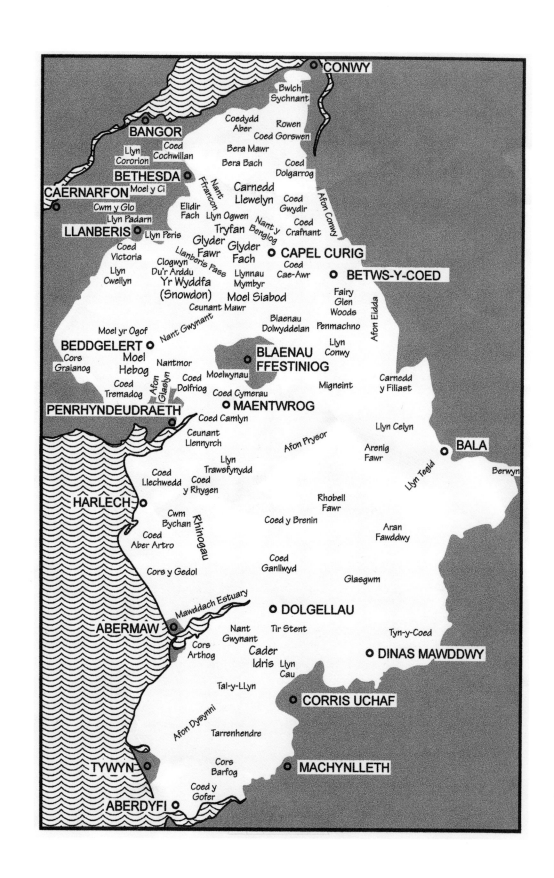

THE PLANT LIFE

OF

SNOWDONIA

INCLUDING THE FUNGI AND LICHENS

Edited by

Peter Rhind and David Evans

First Impression—2001

ISBN 1 84323 044 5

© Text: The Contributors
© Photographs: Peter Rhind, David Evans and Charles Aron (see p. 151 for full details)
© Map: Alan Bradley

Printed in Wales at
Gomer Press, Llandysul, Ceredigion

CONTENTS

PREFACE   vii
by Dick Roberts

ACKNOWLEDGEMENTS   ix

INTRODUCTION   xi
by Peter Rhind

SEED PLANTS   1
by Barbara Jones

FERNS   37
by Dewi Jones

CLUBMOSSES AND QUILLWORTS   56
by Peter Rhind

HORSETAILS   64
by Peter Rhind

MOSSES AND LIVERWORTS   70
by Tim Blackstock and Marcus Yeo

FUNGI   95
by Charles Aron

LICHENS   118
by Allan Pentecost

FRESHWATER ALGAE   137
by Allan Pentecost and Christine M. Happey-Wood

NOTES ON CONTRIBUTORS   149

ILLUSTRATION CREDITS   151

INDEX   153

PREFACE

It sometimes happens that one comes across a book which influences one much more than any other book before. In my case it was not the whole book but a single chapter headed 'Notes on the Flora of Snowdonia', in the 2nd edition of Carr & Lister's classic old book *The Mountains of Snowdonia* (1948). The chapter had been written by J. Bretland Farmer who, it was clear, had researched his subject thoroughly. It was the first time that I realised how rich and how interesting was the plant life of my native land; and how many of the plants I had never seen. I use the words 'native land' deliberately, for I had grown up, many years before, on a farm, over 600 ft above sea level and close to the foothills of the Carneddau.

All the mountain land as far as the watershed along the tops of Yr Aryg, Foel Grach, Carnedd Llewelyn and Carnedd Dafydd was common land on which my family had grazing rights. Here every summer our sheep were taken to graze on the mountain pasture, and during my boyhood I often went with my father when he took them to Yr Aryg. There they stayed on the south-facing slopes until the end of October, except for a few days at the end of June, when they were brought down to be sheared. It was on one such occasion that, on the way home, we made a detour onto the slopes of Moel Wnion, where I first saw the Crowberry with small purplish-black berries; and not far from it a small patch of Bilberry.

Over the years I became acquainted with other mountain plants: the Cowberry among the summit rocks of Bera Bach and Bera Mawr; the Stag's-horn Clubmoss in many places, often gathered by shepherd boys to adorn their caps; the Parsley Fern, almost always among a cluster of stones; Hare's-tail Cotton-grass; and Heath Rush, which my father called 'Troell-gorun', its very apt Welsh name so easy to remember.

But at last, here I was, having spent twelve years (including four in the RAF) on the wrong side of Offa's Dyke; and, within a few months of my return, reading this inspiring chapter in my new home in Betws-y-Coed. It was not long before I had seen the Snowdon Lily (*Lloydia serotina*) – a single plant within a few yards of the path close to the Devil's Kitchen – and before I left Betws-y-Coed for Bangor in 1955, I had seen most of the noteworthy plants in the Snowdon area, and many besides.

And now here is another book, beautifully illustrated and well written, devoted entirely to the more interesting plants, and covering a much larger area than our old 'Snowdonia'. Once again old memories come flooding back: I recall the ferns I saw near the top of the Aber Falls; and the very similar ones on the scree above Llyn Eigiau a few weeks later. None of them could I name, but they showed me the way to the solution of a long-standing problem . . . From then on no one had any difficulty in identifying the Mountain Male-fern (*Dryopteris oreades*).

Best of all perhaps, I like to recall the walk to Cwm Eigiau when, in my haste, I slipped and fell on a damp patch of grass. When I got to my feet, I was holding a handful of muddy grass, among it a piece of a slender plant with white flowers. By sheer luck I had fallen on the Limestone Bedstraw (*Galium sterneri*)!

... By the time I got back to the mountain gate my trousers had dried; and there was my wife, sitting in the car, waiting to take me home over the last few miles.

Happy memories of times long ago, and I hope I shall have many more when I browse through this book again.

My hearty thanks to all who have worked so hard to produce it.

R. H. (Dick) Roberts

ACKNOWLEDGEMENTS

The editors and authors would like to acknowledge with thanks the help of the Countryside Council for Wales and the Snowdonia National Park Authority. The people who have kindly provided assistance include Dr David Allen, Nigel Brown, Ann Giangiordano, Nick Hodgetts, Dr George Hutchinson, Dr Bruce Ing, Dyfed Jones, Dr Peter Jones, Dr Jim Latham, Dick Roberts, Hywel Roberts, Anne Seddon, Dr David Stevens, Alex Turner and Ray Woods.

INTRODUCTION

PETER RHIND

Snowdonia (Eryri in Welsh) is one of Britain's most important areas for plants and other wildlife, and few other natural areas have such a variety of habitats. Its geographical boundaries have never been clearly defined, but since the creation of the Snowdonia National Park in 1951, it has come to be regarded as synonymous with the National Park. On a north-south axis it stretches 70 km from Afon Conwy (the River Conwy) on the north coast of Wales to the northern border of the Dyfi Estuary and, apart from Snowdon itself (Yr Wyddfa in Welsh), includes at least eight other distinct mountain blocks. The Carneddau, the Glyderau, Moel Siabod, Moel Hebog, the Moelwyn, the Arenigs, the Rhinogau and Cadair Idris are all substantial mountains in their own right, and each has its own particular character.

Prior to the establishment of the National Park, the name Snowdonia was more closely associated with the mountains north of Beddgelert in the old county of Caernarvonshire, and the Aberglaslyn Pass, just south of Beddgelert was regarded as the gateway to Snowdonia. However, this area is better known to the Welsh as Eryri, a name possibly derived from the word *eyrie* or breeding place of eagles, although according to another interpretation it may simply mean 'mountain land'. The chapters that follow are based on the wider, National Park, definition, but with the addition of one range of mountains, the Berwyn, that lies just beyond the boundary.

Much of the area is of outstanding conservation interest, including some ninety Sites of Special Scientific Interest and sixteen National Nature Reserves. In addition, habitats such as alpine and sub-alpine heaths, alpine and boreal grasslands, acidic screes and rock crevice vegetation, oceanic woodlands and lakes are now recognised to be of international importance, and under the European Habitats and Species Directive, parts of Snowdonia, including Snowdon, the Glyderau, the Carneddau, the Rhinogau and Llyn Cwellyn have been designated European Special Areas of Conservation. Two of Snowdonia's lakes, Llyn Tegid and Cwm Idwal, have also been recognised as internationally important wetlands under the terms of the Ramsar Convention.

Snowdonia's ecological diversity is partly attributable to its complex geological history. The mountains are composed of some of the oldest rocks on Earth, and because a number of early pioneer studies in geology were carried out in Snowdonia, some of the geological periods bear Welsh names. The oldest rocks date from the Cambrian Period (Cambria being the Latin name for Wales), which started some 600 million years ago when higher life forms were just beginning to evolve. Much of the remaining rock is of Ordovician age (named after the Ordovices, an ancient Welsh tribe), formed at a time some 500 million years ago when the earliest land plants were emerging. Rocks of Cambrian age include the

Rhinogau, while Snowdon is one of the many peaks composed of Ordovician rock. The park also has examples of rocks from the succeeding Silurian Period (again named after an ancient Welsh tribe, the Silures), formed some 430 million years ago when some of the earliest land animals were emerging.

Throughout this immense span of time the rocks of Snowdonia have undergone many changes as a result of exposure to high temperatures and pressures associated with crustal movements and volcanic activity, and on at least two occasions the whole area has been covered by seawater. The present landmass came into existence towards the end of the Silurian Period, but millions of years were to pass before the landform would bear any resemblance to the mountains that we know today. In fact, Snowdonia's mountains are but the weathered remnants of a far larger system, probably much like the Himalayan range in its early geological history. Chemical weathering by atmospheric acids, erosion through wind and water and repeated cycles of freeze and thaw have, over eons of time, conspired to demolish much of the former bulk of these mountains. On occasion, the erosive power of water has been aided by its mass conversion to snow and ice, since during the last two and half million years much of Britain is thought to have experienced at least 50 cycles of glacial advance and retreat. The most recent advance, usually known as the 'The Ice Age', began some 120,000 years ago, and with only a few slightly milder interludes, lasted until about 10,000 years ago. In Snowdonia a huge ice cap developed in the Trawsfynydd area (the Merioneth Ice Cap), and many glaciers formed in the surrounding valleys, some becoming deep 'rivers' of ice extending well beyond present-day river estuaries. One such glacier, in Nant Ffrancon, is thought to have been over 300 m deep and 17 km in length, extending well into the area that is now Ynys Môn (Anglesey).

Although the basic structure of Snowdonia existed before the Ice Age, the massive erosive power of the many glaciers was responsible for some spectacular remodelling. Many of the V-shaped river valleys were converted to deep U-shaped gorges, peaks were sharpened as their flanks were ground away, while below the summits, characteristic amphitheatre-shaped cwms were formed, such as Cwm Idwal. The net effect has been to create some Britain's most spectacular mountain scenery, bearing abundant evidence of glacial action in the form of cirques (cwms), arêtes, troughs and *roche montonnées*, along with glacial depositional features, such as moraines and protalus ramparts. The influence of peri-glacial conditions in which ice thaws and re-freezes seasonally is also apparent in structures such as block screes (sometimes, as in Cwm Idwal, including blocks as big as houses), and extensive block fields such as those on Bera Mawr and Bera Bach in the Carneddau. The frost shattered summits and tors of some mountain summits – as on the Glyderau for example – are also a product of these conditions.

With their complex history, it is not surprising that the rocks show enormous variation in structure, composition and texture. Among the sedimentary rocks – formed by the slow accumulation of sediments on ancient (Palaeozoic) sea floors, are mudstones, sandstones and conglomerates. In some places, such as

Cwm Idwal in winter.

Cwm Graianog, it is still possible to see perfectly fossilized ripple-marks that formed on a Cambrian sea-floor. During the mountain building phase some of these sediments were exposed to such extreme pressures and temperatures that they were modified physically and chemically to form metamorphic rocks. Extensive areas of sedimentary rocks in Snowdonia have been altered in this way, creating, for example, the many outcrops of slate which have played such an important role in the economy of North Wales. There are also many extrusive and intrusive igneous rocks, formed by volcanic action during the Ordovician Period, when the area was still under the sea. The summits of mountains like Cadair Idris and Snowdon are the remnants of such intrusions, protected by their hardness from the erosion that has stripped away the softer surrounding rocks.

The chemistry and physical structure of the rocks has an important influence on the vegetation that grows on them. Calcareous or base-rich rocks such as dolerite and the pyroclastic schists are generally far richer in species than acidic ones such as granite and rhyolite, and provide a habitat for many of the rarities, including the arctic-alpine species. By comparison, the plant communities of acid rocks are species-poor, with almost all the cover coming from a few highly adapted species that can grow only on lime-free soils. Vast tracts of the uplands are consequently dominated by communities in which species such as heather, Bilberry, Mat Grass, Purple Moor-grass and Cotton-grass play a leading role.

Since the most recent ice sheets began to melt some 10,000 years ago, the climate of Snowdonia has gone through various changes, and studies of the pollen and spores in cores from lake sediments have given us an insight into how the vegetation has been altered by these changes. Cores from sites such as Cwm Cywion, Llyn Llydaw, Cwm Idwal, Cwm Melynllyn, Cwm Dwythwch, Clogwyn y Garreg, Llyn Cororion, Cors Geuallt and Nant Ffrancon tell a story of at least five distinct climatic phases. Immediately after the ice retreated, in the so-called Pre-Boreal period (10,000 to 9000 years ago), the climate was still too harsh for all but the hardiest plants, and the vegetation seems to have been a kind of arctic tundra. Further warming, though, signalled the start of the Boreal period (9000 to 7000 years ago) in which a rise in temperature allowed trees such as Birch, Pine and Hazel to flourish. Also warm but with higher rainfall was the succeeding Atlantic Period (7000 to 5000 years ago), when conditions were right for Oak and Lime, and for valley mires to form. The Sub-Boreal Period (5000 to 2800 years ago) marked a return to warm and dry conditions, and then in the final phase, known as the Sub-Atlantic Period (2800 ago to the present) the climate became essentially similar to the one we have today.

Changes during both the Sub-Boreal and Sub-Atlantic periods were heavily influenced by the presence of man, culminating in the widespread clearance of forests and their replacement by cereal crops or grazing pastures. However, since temperature typically falls by about one degree centigrade for each 100 m of elevation, the climate in mountainous regions such as Snowdonia is never uniform. Today it is possible to move through several different climatic zones ranging from a mild temperate climate in the valley bottoms to an arctic-alpine climate in the higher uplands. The extent of these zones is greatly influenced by the climate as whole. During the immediate postglacial period arctic-alpine conditions would have prevailed throughout much of Snowdonia irrespective of altitude, but as the climate ameliorated during the relatively warm Boreal Period, this zone would have become confined to the high summits. Below this there would have been a broad zone of sub-alpine heaths and grasslands, with tree cover spreading up the mountain slopes, eventually reaching about 800 m above sea level. The present slightly cooler phase of the Sub-Atlantic climate has allowed the arctic-alpine zone to reclaim some of its former ground, and at the same time reduced the area available for sub-alpine grass and heather communities. Fully natural development is prevented, though, by the almost universal presence of sheep, and when stocking densities are high, heather communities tend to be replaced by grassland. Woodland is also often prevented from developing by grazing and although the natural tree line is now thought to be at about 470 m, it rarely occurs at this altitude.

The botanical richness of Snowdonia therefore resides to a great extent in the habitats and species that are inaccessible to sheep, or unpalatable to them. Many of the arctic-alpines, for example, are restricted to cliff ledges and crevices, while mosses and lichens survive because sheep have no interest in them as food. In the case of the surviving ancient woodland, species richness is the product of a cool moist oceanic climate persisting over periods of hundreds of years under a continuously closed tree canopy. Any recently planted woodland, or even natural woodland that has been managed for timber is likely to lack the full range of species.

Plant life makes up much of the visible landscape in Snowdonia, with expanses of bracken, grassland, heathland, moorland, wetland or woodland all contributing enormously to the charm and fascination of the area. It was no doubt this appeal and power of the landscape that inspired an unknown person to carve the following poem by Thomas Gray on a rock in Coed Ganllwyd in the 18th century.

> *Oh thou the spirit mid these scenes abiding*
> *Whate'er the name by which thy power be knowing*
> *(Truly no mean divinity presiding*
> *These native streams, these ancient forests own;*
> *And here on pathless rock or mountain height,*
> *Amid the torrent's ever-echoing road,*
> *The headlong cliff, the wood's eternal night,*
> *We feel the godhead's awful presence more*
> *Than if resplendent neath the cedar beam,*
> *By Phidias wrought, his golden image rose),*
> *If meet the homage of thy vot'ry seem*
> *Grant to my youth, my weary youth, repose.*

SEED PLANTS

BARBARA JONES

INTRODUCTION

Development of the seed was one of the most important innovations in the evolution of the vascular plants, and is one of the main factors responsible for the dominance of seed plants in today's flora. In these plants all stages of embryonic development take place within the seed, so that there is no free living sexual or gametophyte generation as in the ferns and other lower plants.

The earliest seed plants, the gymnosperms, first appeared in late Devonian times some 350 million years ago. The name means 'naked seed', and refers to the fact that, unlike in the flowering plants (see below), seeds are borne externally. The living representatives of the group include the conifers, which date back to at least the Carboniferous Period, some 290 million years ago. Before man's introduction of various exotic conifers, Juniper (*Juniperus communis*) and possibly Yew (*Taxus baccata*) may have been the only conifer species in Snowdonia.

By far the largest group of seed plants however, is the flowering plants or angiosperms. They differ from the gymnosperms in having their seeds enclosed by a carpel, and are relative newcomers to the plant kingdom, with no fossil record until Cretaceous times some 125 million years ago. Within 25 million years, though, they had become the dominant plant group on Earth, and today make up much of the visible world of plants. Much of this success can be attributed to the evolution of the flower, which has enabled them to use wind, insects and other animals in the transfer of reproductive cells or pollen from one plant to another. In this respect, they have transcended their rooted condition and achieved mobility almost analogous to that of animals.

However, plants that rely on annual seed production and germination for their survival are unlikely to do well in the unpredictable often harsh conditions of mountain environments. Mountain plants therefore tend to be long-lived perennials with an ability to proliferate by methods other than seed production. The heathers for example, undergo slow growth of additional stems and leaves; certain grasses and sedges spread by underground rhizomes, while other species spread by release of plant parts. Some saxifrages, for example can reproduce by means of small bulb like structures (bulbils), which are produced above ground in some species. These eventually fall to the ground and produce new plants. A strategy adopted by a number of mountain grasses such as Alpine Meadow-grass (*Poa alpina*), Alpine Tufted Hair-grass (*Deschampsia cespitosa* ssp. *alpina*) and Viviparous Fescue (*Festuca vivipara*) is to circumvent the risky business of seed production and germination by becoming viviparous. Instead of seeds, these plants produce small plantlets, which drop off when they are capable of taking root. The unreliability of pollination due to adverse weather or shortage of pollinators in the mountains has also resulted in adaptations to self-pollination in several species. This is achieved when the anthers bend over to deposit pollen on to the stigma of the same flower. It can happen in an open flower – perhaps as a fallback when cross pollination fails (Snowdon Lily (*Lloydia serotina*) and Tufted Saxifrage (*Saxifraga cespitosa*) are examples) – or it may be obligate where flowers never open their petals (as in the eyebright *Euphrasia cambrica*). Such strategies are not usually the sole means of reproduction, however, and many mountain plants can employ either sexual or vegetative methods according to circumstances.

THE PLANT HUNTERS OF SNOWDONIA

Plant hunting has taken place in Wales for hundreds if not thousands of years. The earliest uses of plants would have been for religious and ceremonial purposes by the Druids, with magical or healing powers ascribed to many of them. These healing powers were later recognised and developed in the Middle Ages by physicians, many of them in monasteries ministering to local people, but it wasn't until the 16th century, helped by the publication of *A New Herbal* (William Turner 1551), that the study of plants really became a pursuit in its own right. Although this was published in England, it greatly influenced the Welsh herbalist the Rev. William Salesbury who was based at Plas Isaf near Llanrwst. Salesbury published his famous Welsh herbal in the 16th century and, although the original manuscript was lost, a copy produced in 1763 survives in the National Library of Wales. An edited version published by Roberts in 1916 also exists. These works still concentrated on the medicinal value of plants but, for the first time, there were references to plant localities.

The famous English herbalist John Gerard published his *Herbal or General Historie of Plants* in 1597, but more relevant to Wales was a second edition which mentions several plants found in Wales. The herbal was updated and subsequently published in 1633 by Thomas Johnson, an apothecary from London, and includes the first published list of plant names in Welsh. Robert Davyes, an antiquarian from Flintshire, had supplied the plants. Johnson was quite an explorer for his day, and when he learned that Wales contained a number of plants that he was unlikely to see in England, he resolved to visit the country. On arriving in Snowdonia in 1639, he lost no time in mounting an expedition to the mountains, which he called the 'British Alps'. On Snowdon he found a number of interesting plants including Alpine Saw-wort (*Saussurea alpina*), Dwarf Willow (*Salix herbacea*), Moss Campion (*Silene acaulis*), Mountain Sorrel (*Oxyria digyna*), three saxifrages and, to his surprise, far from the sea, Sea Campion (*Silene uniflora*) and Sea Pink (*Armeria maritima*). His visit to Ysgolion Duon on Carnedd Llewelyn was less successful, though, in part at least because, according to Johnson, his guide decided to turn back for fear of eagles.

Thrift (*Armeria maritima*) in Clogwyn Du'r Arddu.

The next important plant hunter to visit Snowdonia was John Ray, a university lecturer from Cambridge whose interest in plants went beyond their medicinal value. On his first visit in 1658 he had to curtail an ascent of Snowdon due to bad weather, but he later climbed Cadair Idris, where the only new flowering plant he found was Globe-flower (*Trollius europaeus*). Four years later, in 1662, he returned to Wales, and made an ascent of Carnedd Llewelyn. According to his account though he '. . . had not time enough to search the rocks, and so found no rare plants there, only *Saxifraga stellaris* [Starry Saxifrage] . . .'. However, he later travelled to Llanberis and found the Welsh Poppy (*Meconopsis cambrica*) growing near Llyn Peris, and the now very rare Small White Orchid (*Pseudorchis albida*) near an old stone wall. Ray never returned on Wales after 1662, but went on to publish a number of important treatises on British plants. Most of his Welsh records appear in either *Fasciculus Stirpium Britannicarum* (1688) or *Synopsis Methodica Stirpium Britannicarum* (1690).

The greatest additions to the Welsh flora came from the work of the outstanding Welsh botanist Edward Lhuyd. He was born near Oswestry and had a range of interests and expertise that included chemistry, archaeology,

Mountain Sorrel (*Oxyria digyna*) in Cwm Idwal.

palaeontology, natural history, philology and botany. As keeper of the Ashmolean Museum at Oxford, he travelled widely in search of material for the Museum, and we know from Ray's books that about fifty of the plants he recorded were Welsh 'firsts'. His name is best remembered, though, for his discovery of the Snowdon Lily (*Lloydia serotina*) in Wales. It was initially placed in the genus *Anthericum*, but later investigations showed that it merited a genus of its own, and the name *Lloydia* was chosen in recognition of Lhuyd's achievement. When he first saw the plant it was not in flower and he described it as 'A certain rush leaved, bulbous plant, having one seeded vessel on the top of an erect stalk about nine inches high. It hath three or four more narrow and short leaves upon the stalk'. Interest in the plant grew, and a number of eminent botanists of the 18th and 19th centuries mounted expeditions to find specimens, which, according to Richard Richardson, were 'springing out of the naked rocks'. Early references indicate that its distribution in the early days was more or less the same as today, but there is no doubt that past collecting and heavy sheep grazing have taken their toll. It appears to have been difficult to find and collect even in the 18th century, as indicated by references from these early expeditions. Richardson wrote '. . . it is a very hard matter for anybody to find it that does not know the particular place exactly,' and R. A. Salisbury in his 1812 paper reported that Dr William Alexander of Halifax '. . . was near losing his life in climbing to the dangerous summits where it grows wild.' Nevertheless, large numbers of plants were collected, with the result that some locations now hold very few plants, and most sites are accessible only by rope.

Mossy Saxifrage (*Saxifraga hypnoides*) in Cwm Idwal.

From the very early days of botanical exploration, mountain plants have been a focus of interest in Snowdonia, and visiting plant hunters often used local guides to track down the rarer species. In the 17th century, the well-known botanist Richard Richardson, for example, often used guides during his exploration of Snowdonia. He was a close friend of Edward Lhuyd, and after Lhuyd's death in 1709, he became the leading authority on Welsh mountain plants. It is interesting to note that in the copy of Ray's second edition of the *Synopsis Methodica Stirpium Britannicarum* (1696) kept in the National Museum of Wales, a hand-written note by Richardson describes how Mr Lhuyd had shown him the Snowdon Lily on the side of Trigfylchau towards Nant Ffrancon. Trigfylchau is a locality mentioned by several botanical explorers during the 17th and 18th centuries, but the name has gone out of use and all we now know for certain is that it was a spur of Glyder Fawr.

In 1726 an expedition by the German botanist John Jacob Dillenius and Samuel Brewer to explore Cwm Glas Mawr and Clogwyn y Garnedd on Snowdon and Cadair Idris utilised instructions provided by Richardson. On the latter they found a number of interesting plants, including Mossy Saxifrage (*Saxifraga hypnoides*), Starry Saxifrage (*S. stellaris*) and Roseroot (*Sedum rosea*), while at the base of the mountain they recorded Round-leaved Sundew (*Drosera rotundifolia*) and Long-leaved Sundew (*D. intermedia*). In Cwm Glas Mawr they noted Mountain Sorrel (*Oxyria digyna*) and in the small mountain lake called Llyn Glas saw Water Lobelia (*Lobelia dortmanna*) and Shore-weed (*Littorella uniflora*). They continued to climb towards the summit of Snowdon, and among the wet rocks of Crib Goch found Alpine Meadow-rue (*Thalictrum alpinum*) and Moss Campion (*Silene acaulis*).

The weather often appears to have been dreary on these early expeditions, but the rewards clearly outweighed any discomfort they may have experienced. In fact, they were great enough to entice Brewer to remain in North Wales for a further twelve months, and make a number of excursions, including 13 ascents of Snowdon. During this period Brewer kept a diary, and although the original was lost, two transcripts were made. One is now in the British Museum (Natural History), and the other in the National Museum of Wales, Cardiff. The diary shows that he made several visits to Trigfylchau where he found species such as Alpine Saw-wort (*Saussurea alpina*), Globe-flower (*Trollius europaeus*) and Purple Saxifrage (*Saxifraga oppositifolia*). However, Cwm Glas Mawr was his favourite site, and it is still a productive hunting ground for mountain plants.

In 1773 Sir Joseph Banks and his friend the Rev. John Lightfoot visited Snowdonia during a major plant-hunting expedition through Wales. Banks by then was famous for his travels with Captain Cook aboard HMS *Endeavour* and the plants he had collected from many parts of the world eventually formed the basis of the Herbarium at the British Museum (Natural History). Many of the best alpine sites in Snowdonia were already known by this time, but Banks was keen to follow in the footsteps of the old master botanists such as John Ray. Nevertheless, rarities such as the Snowdon Lily still eluded many of these expert botanists, including both Banks and Lightfoot on their visit to Snowdon. Lightfoot did, however, find Alpine Meadow-grass (*Poa alpina*), Alpine Saw-wort and Roseroot (*Sedum rosea*) on Clogwyn y Garnedd, and in a joint excursion to Ysgolion Duon they found the rare Three-flowered Rush (*Juncus triglumis*).

One of the first books to deal with the distribution of plants was *The Botanist's Guide through England and Wales* published in 1805 by D. Turner and L.W. Dillwyn, and it listed the uncommon and rare plants that had been recorded from each of the English and Welsh counties. By this time Caernarvonshire – which included Snowdonia – was one of the best-recorded counties, although Turner and Dillwyn still thought that the mountains of Snowdonia 'afford inexhaustible fields for discovery.' It was not until 1894 that *The Flora of Caernarvonshire and Anglesey* was published. Its author, J. E. Griffith, a former Bangor chemist, had brought together everything that was then known about the flowering plants, ferns and lower plants of the area, making it one of the most complete floras ever produced. There has been nothing comparable since then, although a number of papers dealing with parts of Snowdonia have been published. In the 1920s N. Woodhead described the flora of Llyn Ogwen and Llyn Idwal, and A. Wilson provided a description of the plants found on Tal-y-Fan. Both appeared in the 1927-1928 *Proceedings of the Llandudno, Colwyn Bay and District Field Club*.

Like many of Snowdonia's lakes, the acidic waters of Llyn Ogwen and Llyn Idwal included species such as Water Lobelia (*Lobelia dortmanna*) and Awlwort (*Subularia aquatica*). Tal-y-Fan, which forms a north-eastern spur of the Carnedd Llewelyn range, was not found to be particularly rich in flowering plants, but the uncommon Ivy-leaved Bellflower (*Wahlenbergia hederacea*) and Mountain Everlasting (*Antennaria dioica*) were recorded. In 1940s, A. Wilson went on to publish *The flora of a portion of north-east Caernarvonshire* in the *North Western Naturalist* (1946/1947). This area, which included the entire Carneddau range, was also described as rather poor compared with Snowdon and the Glyderau, but he still managed to record 14 arctic-alpine species including Mountain Avens (*Dryas octopetala*) on the hills east of Llyn Cowlyd, and Chickweed Willowherb (*Epilobium alsinifolium*) on Foel Grach.

Further contributions to our knowledge came in the 1960s from four local botanists, Dick Roberts, Evan Roberts, Mary Richards and Peter Benoit. In 1961, Benoit and Richards published *A Contribution to the Flora of Merioneth* in the natural history journal *Nature in Wales*. In this southern part of Snowdonia, they recorded about 23 'highland' species, but described the alpine flora as poor compared with that of neighbouring Caernarvonshire. The publication of *Plant notes from southeast Caernarvonshire* in 1963 by Evan Roberts and Dick Roberts in the *Proceedings of the Botanical Society of the British Isles* complemented Wilson's 1940s flora for northeastern Caernarvonshire. The area, which includes Moel Siabod, was described as mainly composed of acidic terrain and therefore of little botanical interest, but lime-rich rocks of botanical interest were listed for several areas. They mentioned in particular the calcareous volcanic tuffs in the Blaenau Dolwyddelan–Nantgwynant area, and between Capel Curig and Dyffryn Crafnant. They recorded several uncommon species including Lesser Twayblade (*Listera cordata*) near Capel Curig, and several other orchids, but they were concerned that a population of the Bog Orchid (*Hammarbya paludosa*), recorded in a small bog near Bryn Cathlwyd in 1946, seemed to have been destroyed by land drainage.

In more recent years, botanical studies in Snowdonia have ranged from description of vegetation communities to detailed research on individual species. Records of plant communities are valuable in providing a baseline against which change can be measured. One such 1950s study by Derek Ratcliffe on the Carneddau provided data that are now proving useful in assessing the effects on wild plants of agriculture, recreation, acid precipitation and climate change. Although we are probably past the age of discovery of new species in Snowdonia, there is still a wealth of information to be gathered from studying plants in their habitats, and there is still a good chance that new localities for rare species will be discovered.

IMPORTANT HABITATS FOR SEED PLANTS IN SNOWDONIA

Seed plants and especially the flowering plants dominate many of the habitats of Snowdonia, and some of the more important ones are discussed in the following sections. Although the flora of Snowdonia could be considered poor in species, compared to the mountains of Europe or the Scottish Highlands, it has its own special interest in its unusual diversity of habitats reflecting arctic, alpine, maritime and lowland influences within a relatively small area.

MOUNTAIN TOPS

Extremes of wind and temperature on the summits of the higher mountains in Snowdonia mean that few seed bearing plants can survive. Nevertheless, species such as the Stiff Sedge (*Carex bigelowii*) and the hardy Dwarf Willow (*Salix herbacea*) are virtually confined to high ground. The latter species is locally common in montane heath communities that are not too heavily grazed. The small size of its stems and leaves initially give the impression of a low growing Bilberry, but closer inspection reveals it to be a tree in miniature. Also in these mountain top communities are small tussocky grasses, especially Sheep's Fescue (*Festuca ovina*), Velvet Bent (*Agrostis canina*), Viviparous Fescue (*Festuca vivipara*) and Wavy Hair-grass (*Deschampsia flexuosa*), but none reaches the size or luxuriance that it would achieve at lower altitudes. The summits of the Glyderau and the Carneddau are two of the best places to see these communities.

Castell y Gwynt on the summit of Glyder Fach.

ROCKY LEDGE COMMUNITIES

Rocky ledges are an important habitat for upland plants in Snowdonia.Their inaccessability to sheep makes them a refuge for many plants with sensitivity to grazing, particularly tall growing herbs and ericaceous shrubs. Sites with base-rich soil or that have mineral-rich water flushing through them are especially rich. The description 'tall-herb ledges' indicates their nature, and in summer they can provide some of the most attractive and spectacular displays of wild plants in the mountains, often with the appearance of hanging gardens. Some of the shaded sites create conditions suitable for species that would normally be associated with woodland. Indeed, one of the most characteristic species of this community is Wood-rush (*Luzula sylvatica*) which can grow luxuriantly on the cool, humid ledges, often in company with other typical woodland species such as Angelica (*Angelica sylvestris*), Golden-rod (*Solidago virgaurea*), Water Avens (*Geum rivale*) and Wood Anemone (*Anemone nemorosa*). Some of the best examples can be seen on the crags of Twll Du in Cwm Idwal, Clogwyn y Garnedd on Snowdon, and on the precipitous cliffs of Ysgolion Duon at the head of Cwm Llafar on the Carneddau. In these places one can also see several of the less common herbs such as the Globe-flower (*Trollius europaeus*) – a member of the buttercup family which grows in a number of northern montane habitats, Lady's Mantle (*Alchemilla glabra*) and Welsh Poppy (*Meconopsis cambrica*).

Welsh Poppy (*Meconopsis cambrica*) in Dyffryn Ogwen.

In fact, Twll Du is one of the few places where Welsh Poppy can be regarded as truly wild – although it is naturalised in many places at lower altitudes. It is possible, though, that this was not always the case, and when the eminent botanist John Ray saw it at Llyn Peris in 1662, it may well have been a genuine wild plant. Roseroot (*Sedum rosea*) is also often seen on these ledges and can form luxuriant masses of greenish-yellow flowers extending from the ledges on to the surrounding rock faces. Hugh Davies used the Welsh name for this plant, 'Pren y Ddanodd', referring to its former use as a cure for toothache, although how such knowledge was originally acquired is hard to imagine.

Roseroot (*Sedum rosea*) in Cwm Idwal.

On his visits to Carnedd Llewelyn and Snowdon, Ray would have also seen other members of the tall-herb ledge community, including Mountain Sorrel (*Oxyria digyna*), Alpine Meadow-rue (*Thalictrum alpinum*), Lesser Meadow-rue (*Thalictrum minus*), and possibly Harebell (*Campanula rotundifolia*). The striking pink flowers of Thrift (*Armeria maritima*) might also have been seen on a spring visit. Widely distributed in the northern hemisphere, Thrift is usually a coastal plant, but like several other species, it can be found in both maritime and mountain situations. Scurvy-grass (*Cochlearia officinalis*) is another example, notable for the fact that its leaves, rich in vitamin C, were once valued by early seafarers as a protection against scurvy. What is less well known is that shepherds have been known to use them in Snowdonia, perhaps for the same purpose.

Contributing to the ephemeral but occasionally vivid colours of the ledge communities are a number of more typical lowland plants. Ox-eye Daisy (*Leucanthemum vulgare*), Meadow Buttercup (*Ranunculus acris*), and Devil's-bit Scabious (*Succisa pratensis*) are examples and, intermingled with them, the viviparous forms of Fescue (*Festuca vivipara*) and Tufted Hair-grass (*Deschampsia cespitosa* ssp. *alpina*).

ARCTIC-ALPINE SPECIES OF ROCKY OUTCROPS

There is little doubt that many people are drawn to Snowdonia to see the arctic-alpine plants. These 'jewels' of the mountains usually grow on high mountain crags well out of the reach of grazing animals, and it seems remarkable that such delicate looking plants can survive year after year in some of the most hostile conditions in Britain. In fact, though, they are well adapted to the habitat, and under gentler conditions would soon be overcome by competition from more aggressive species. Snowdonia is not as richly endowed with these native mountain species as the Alps, the mountains of Scandinavia, or even the Scottish Highlands, but the fact that they are on the edge of their European distribution gives them special significance in our understanding the postglacial history of the British Isles. Snowdonia has been a refuge for a number of species that must have first established themselves on high ground as the ice retreated. They are true arctic species, and have survived to the present day only by colonising the coldest sites available – on high altitude north-facing cliffs with no direct sunlight. The lack of such terrain to the south means that Snowdonia is the most southerly British outpost of many species. The Tufted Saxifrage (*Saxifraga cespitosa*), for example, with a wide circumpolar distribution

Tufted Saxifrage (*Saxifraga cespitosa*) in Cwm Idwal.

including Alaska, Arctic Canada, Greenland, Scandinavia and Siberia, is limited in Britain to a few small colonies, mainly in Scotland, and to altitudes above 500 m. Our single remnant population in Snowdonia is one of the most southerly in Europe. In its arctic environment, it is a primary coloniser of moraines and recently exposed ground, and can survive and reproduce in very harsh environments. In Britain it occurs almost always on moderately base-rich, north-facing rocks, with few other flowering plants and bryophytes. The pollen record suggests that it was widespread in the British Isles immediately after the last Ice Age, but became confined to isolated upland sites as the climate warmed and competition from other plants in the more equable lowlands increased. There is also the possibility that Tufted Saxifrage and other arctic-alpine species may have survived the Ice Age in Snowdonia and Scotland on ice-free rock outcrops known as nunataks.

By 1977 the Tufted Saxifrage population in Snowdonia had declined to four small plants, bringing it to the brink of extinction in the area. While the population had probably always been small – because of the scarcity of its habitat – collecting over the past 200 years must also have taken a toll, and there was an attempt in the late 1970s to boost plant numbers artificially. Hundreds of plants were grown from seeds taken from one of the remaining wild plants, and seeds, seedlings and mature plants reintroduced. Their fate is still being monitored today, and although the experiment was partially successful, numbers remain low and very few seeds seem to germinate naturally. A possible conclusion is that the climate and environment of Snowdonia no longer ideal for the species, and that the human passion for collecting has merely hastened a decline that would have occurred anyway.

The cliffs of Clogwyn y Garnedd and Cwm Glas below the summit of Snowdon, Twll Du in Cwm Idwal and Ysgolion Duon on the north side of Carnedd Dafydd all support important communities of arctic-alpine plants, but they can also be found wherever there are open, base-rich rocks inaccessible to grazing, as on the boulders in Cwm Idwal. They survive by growing in small pockets of soil in cracks and crevices and are generally small and often hug the rock by means of creeping stems, or adopt a cushion shape to minimise their exposure to strong winds.

A species which seems to go against several of the generalisations about arctic-alpines is the Snowdon Lily (*Lloydia serotina*). It has no requirement for base-rich substrates, and with its delicate stems and leaves seems ill-equipped to endure strong winds – unless the principle is to accommodate the wind by moving with it. This little plant is one of Snowdonia's most celebrated, partly because it is one of very few that are not also found in Scotland. It is widespread, though, in other mountainous regions of the northern hemisphere, where it often seems to have different habitat preferences – in some cases growing on gently rolling alpine meadows rather than on cliffs. It has attracted visitors to Snowdonia ever since its discovery in the late 17th century by Edward Lhuyd, and has acquired several local names, including 'Brwynddail y Mynydd' (the rush-leaved mountain plant) and 'Y Bryfedog' (the Spiderwort), both of which describe the long, grass-like leaves and general 'spidery' nature of the plant.

Snowdon Lily (*Lloydia serotina*) in Cwm Idwal.

Snowdonia's Snowdon Lily populations are hundreds of miles from their nearest neighbours in the Alps, raising the question of how the species ever reached Snowdonia, and why it should be found here but not in Scotland. A possible explanation is that when it followed the retreating ice at the end of the last Ice Age, it reached only as far north as Snowdonia before conditions warmed to the point that further northward movement became impossible. Alternatively, it may have survived in Snowdonia, through, at least the later ice advances, on one of the previously described glacial refuges, which may not have been available in the more severe glacial environment of Scotland. Whatever the reasons, the genetic differences between our populations and those of the Alps seem to indicate a very long period of separation. Even the small populations in Snowdonia, which are less than five miles apart, are effectively isolated from each other by intervening deep valleys and heavily sheep grazed mountain slopes.

Today, the Snowdon Lily grows on only six or seven cliffs and, unlike most arctic-alpines, shares its habitat with very few other species. It often grows in lines out of cracks or from under overhangs, the result of an ability to spread by rhizome-like tendrils arising from sub-surface bulbs. Viable seed is produced, but it is not yet known how successful this is as a mode of reproduction in the wild.

Few groups of plants are more at home in mountains than the saxifrages. The name literally means 'rock breaker' but this belies the delicacy and beauty of many of the species. Few summer visitors see the magnificent Purple Saxifrage (*Saxifraga oppositifolia*) in bloom, as it can flower as early as February on some of the boulders in Cwm Idwal and on the lower rocks of Snowdon, and may even flower in snow. It is reputed to grow further north than any other flowering plant, having been recorded at Cape Morris Jessup on the north coast of Greenland. It also has the unusual ability to secrete excess lime from glands on its leaves, so

Purple Saxifrage (*Saxifraga oppositifolia*) in Cwm Idwal.

that it appears to be encrusted with lime. True to its name, the flowers are normally purple, but white varieties also occur. The two most common saxifrages in Snowdonia are Mossy Saxifrage (*Saxifraga hypnoides*) and Starry Saxifrage (*S. stellaris*), both of which occur at relatively low altitudes, and Starry Saxifrage is one of the few arctic-alpines able to grow on acid rocks. The Alpine Saxifrage (*Saxifraga nivalis*) is restricted to damp crevices where competition is low, and it is consequently very scarce. Even at the best sites such as Cwm Idwal and Cwm Glas, it can be hard to find. The Irish Saxifrage (*S. rosacea*) was also a Snowdonian species until a few years ago, but now seems to have been lost. There were sporadic records in the the late 18th century, but none in the 20th century until 1962, when a specimen was collected and placed in a local garden. Unfortunately no other wild plant has been found since then, and as far as the British Isles is concerned, Ireland is the species' only remaining refuge. Most arctic-alpine species are rather rare in Snowdonia, but not always because of plant collecting or grazing: scarcity of habitat can also be an important factor.

Starry Saxifrage (*Saxifraga stellaris*) in Cwm Idwal.

Alpine Saxifrage (*Saxifraga nivalis*) in Cwm Idwal.

A species which as yet appears to be unaffected by collecting and still graces many rock faces with bright pink cushions of flower in spring and early summer, is Moss Campion (*Silene acaulis*). It has been used experimentally in Canada to date recent glacial deposits by relating cushion sizes to known growth rates.

Moss Campion (*Silene acaulis*) in Cwm Idwal.

The fact that it grows very slowly in this enviroment means that dating is possible over a surprisingly long time scale. In high mountains or at high latitudes, this cushion habit of growth is rather common because it lowers wind resistance and improves the plant's retention of metabolic heat. Similar protection is given by a covering of hairs, which helps to reduce heat loss in arctic-alpine species such as the Alpine Mouse-ear (*Cerastium alpinum*), Arctic Mouse-ear (*C. arcticum*) and Mountain Everlasting (*Antennaria dioica*). Unfortunately, such hairiness is not unique to the arctic-alpine mouse-ears, and it is possible to mistake a lowland relative such as Common Mouse-ear (*Cerastium fontanum*) – which also grows in the mountains – for one of the true alpine species.

The early botanical explorer Thomas Johnson first noted Northern Rock-cress (*Arabis petraea*) on Clogwyn Du'r Arddu in 1639, and this remains one of its strongholds, although it also occurs in Cwm Glas and a few other places. Its only other British stations are in north-west England and the Scottish Highlands.

Alpine Saw-wort (*Saussurea alpina*) in Ysgolion Duon.

Other attractive arctic-alpine species growing in Snowdonia are Alpine Cinquefoil (*Potentilla crantzii*), Chickweed Willowherb (*Epilobium alsinifolium*), Alpine Saw-wort (*Saussurea alpina*), Northern Bedstraw (*Galium boreale*), Hoary Whitlowgrass (*Draba incana*) and Mountain Avens (*Dryas octopetala*). Mountain Avens is very rare, however. The name *Dryas* is Greek for oak, and the leaf is indeed similar in shape to an oak leaf, though much smaller.

Mountains Avens (*Dryas octopetala*) on the Glyderau.

Its flower is large and attractive with eight white petals, and an ability to rotate on the stem so that it is constantly facing the sun, thereby maximising its absorption of the sun's heat. This is thought to make the flowers more attractive to pollinating insects. At first sight, the bright yellow flowers of Alpine Cinquefoil could be mistaken for one of its lowland relatives, although its flowers often have a distinctive orange spot on each petal. It reproduces mainly by seed and, although pollination is necessary to stimulate seed production, no actual fertilisation is involved. Although Chickweed Willowherb is found on wet rocky ledges, it is essentially a plant of springs and open stony flushes. It has a wide altitudinal range and reaches its southern UK limit in Snowdonia. It is susceptible to grazing though, which may restrict it to sites that are inaccessible to grazing stock.

The Alpine Saw-wort, a member of the daisy family, has a compact head of purple flowers with a delicate fragrance and leaves whose undersides are covered with short, silvery-white hairs. It generally grows on mountain cliffs and rocky outcrops, but descends to sea level in the north of Scotland – as do a number of our other mountain species, including the Hoary Whitlowgrass and Northern Bedstraw. Hoary Whitlowgrass, recognised by pods that twist when ripe, is confined to Snowdonia in Wales, but again can be found in the lowlands, on sand dunes in north and west Scotland. Northern Bedstraw has whorls of four three-veined leaves and fruits covered in hooked bristles. It grows mainly on rocky and gravelly places in the hills, but also on sand dunes in Scotland. The occurrence of these species, and others, in a maritime situation in northern Scotland reflects the climatic similarities with upland regions well to the south. Another species, which also grows in lowland situations, but away from the coast and further south than northern Scotland, is Spring Sandwort (*Minuartia verna*). Though not strictly an arctic-alpine, being mainly a plant of limestone soils and mineral waste containing heavy metals, it does associate with arctic-alpines in the mountains. It can be found growing on rocky outcrops where its slender tufts of flowers appear to have a tenuous grip on the soil.

The arctic-alpines also include some less showy species such as various grasses, rushes and sedges. Notable grasses are Alpine Meadow-grass (*Poa alpina*), Glaucous Meadow-grass (*Poa glauca*) and Alpine Tufted Hair-grass (*Deschampsia cespitosa* ssp. *alpina*), all of them quite rare. In Norway, where grazing is light, the Alpine Meadow-grass is not confined to cliffs, but grows also in dwarf herb grassland. It occurs in such situations only rarely in Britain, mainly in Scotland, perhaps because of sensitivity to grazing. The Glaucous Meadow-grass has similar habitat preferences, and a similar sparse distribution. As mentioned earlier, some of these upland grasses are viviparous, bearing small plantlets on their inflorescences instead of seeds. This is an adaptation for reproduction under conditions that are too cold and wet for normal pollination and seed ripening. The downside of the strategy is that it demands a continuously moist substrate for the plantlets to develop, though this is never likely to be a limitation in the British uplands. The Alpine Tufted Hair-grass appears to be less restricted to calcareous substrates than many arctic-alpines, but competes poorly with other species.

Of all the arctic-alpine sedges and rushes native to the British Isles, only one arctic-alpine rush, the Three-flowered Rush (*Juncus triglumis*), and two arctic-alpine sedges, the Black Alpine-sedge (*Carex atrata*) and Stiff Sedge (*C. bigelowii*), occur in Snowdonia, which is their only Welsh location. The Stiff Sedge is essentially a species of the highest stony mountain heaths. The Three-flowered Rush occurs sporadically on boggy or rocky places mainly on acid soils, whereas the Black Alpine-sedge prefers more calcareous substrates on rock faces or ledges. The best population of the Black Alpine-sedge south of the Scottish Highlands is on Snowdon, but generally the Welsh representation of these arctic-alpine sedges and rushes is poor compared with the populations of the Scottish Highlands. The climate of these Scottish mountains has more in common with the Scandinavian arctic and sub-arctic environments, and is closer to the centre of distribution of these species than the mountain regions in the southern half of Britain. Being so far south, the Snowdonia mountains can support only fragments of the arctic-alpine habitats that are so extensive in the Cairngorms, for example.

There are a number of rare mountain hawkweeds in Snowdonia, including *Hieracium holosericeum* and *H. snowdoniense*. As a group, the hawkweeds have received rather little conservation effort in the past – perhaps because they are often difficult to identify, but there is evidence of a decline in some of the species – again with sheep as likely culprits. Among the species are several endemic or significantly isolated species that are not particularly hard to identify. *H. holosericeum* is a true arctic-alpine, distributed across the mountains of Scotland, Iceland, Scandinavia and central Europe and known from four sites in North Wales (Carnedd Dafydd, Craig yr Ysfa, Tryfan and Glyderau). It appears to have declined on the Glyderau since Brewer in 1727 reported it '. . . on most of the low rocks that surround the lake between the two mountains [Llyn y Cŵn] many seedling plants [of *Hieracium holosericeum*].' The most recent records are from a survey in 2000, which revealed seven patches in two sites. Little is known about the distribution of *H. snowdoniense* except for the sites named in the *Flowering Plants of Wales* (R. G. Ellis, 1983) and these include: Cwm Glas Bach, Cwm Idwal, Nant Ffrancon, Ysgolion Duon and Cwm Perfedd. It appears to be endemic to Snowdonia, and is now listed in the Red Data Book of endangered species.

SCREE SLOPES

Scree slopes are usually composed of angular rocks that have broken away from the rocky cliffs above through a process known as ice wedging, in which repeated cycles of freeze and thaw lead to an opening-up of existing cracks. The word scree is derived from an old Norse word meaning rubble which slides under foot, and it is certainly the case that active screes (with material still being added) are easily disturbed. In Snowdonia these distinctive landforms are widespread, reflecting the geology of their parent cliffs. They can be acidic or basic, or even a mixture of both, and include rocks ranging in size from huge boulders (like those below Twll Du in Cwm Idwal), to fragments that can be disturbed by walking on them (in Cwm Cneifion, for example). Scree vegetation varies depending on stability, and the size and geology of the rocky fragments. The boulder scree below Twll Du, for example, is composed of base-rich volcanic tuffs and acidic rocks, but it is only the tuffs that provide a habitat for the rare arctic-alpines. Many of the base-rich boulders support important communities of arctic-alpine plants, including Alpine Saxifrage, Purple Saxifrage and Spring Sandwort, whereas the acidic boulders are usually devoid of anything other than a few grasses and Heather (*Calluna vulgaris*). Where screes are unstable and acidic (the most common type in Snowdonia), the vegetation is usually composed of a pioneer community of Parsley Fern (*Cryptogramma crispa*) (see the Chapter on Ferns), with various bryophytes, and fine leaved grasses including Wavy Hair-grass and Sheep's Fescue (or Viviparous Fescue at higher altitudes). There may also be patches of Heath Bedstraw (*Galium saxatile*), and grasses such as Mat Grass or Sweet Vernal Grass (*Anthoxanthum odoratum*). Few of the higher plants are common, however, and most can be found in greater abundance elsewhere. Because this type of vegetation is widespread in Snowdonia, with examples in the Llanberis Pass, Nant Ffrancon and the Tal y Llyn Pass, we may assume that it is nothing very special. In fact though, over Europe generally, it is a scarce community.

Where screes have become stabilised at lower altitudes, as in parts of the Nantgwynant, in the Rhinogau and around Llanberis, scrub and woodland have developed, and the shaded rocks and boulders have become covered in a luxuriant carpet of mosses and liverworts. These intensely green, boulder-strewn woodlands are some of the most enchanting places in Snowdonia, especially in the dappled light of early spring. They are generally too dark for flowering plants, but a few may be found in the less shady areas. English Stonecrop (*Sedum anglicum*) and less commonly Rock Stonecrop (*S. forsterianum*) can be found on the tops of boulders such as in the south-facing screes below the Tremadog cliffs near Porthmadog.

MOUNTAIN FLUSHES AND AQUATIC HABITATS

In the uplands, wet hollows, seepage lines or flushes often have interesting assemblages of plants including the now uncommon Bog Orchid (*Hammarbya paludosa*), which reproduces by seed and by means of bulbils which develop on the leaf tips. It often grows in *Sphagnum* moss but can be difficult to find due to its small size and inconsistent flowering from year to year. Associated species in the habitat include Sundew (see below), Cross-leaved Heath (*Erica tetralix*), White Beak-sedge (*Rhynchospora alba*) and occasionally Cranberry (*Vaccinium oxycoccus*).

The shallow waters at the margins of the many small lakes (llynau in Welsh) in Snowdonia are often rich in species, with ferns, horsetails and flowering plants all well represented. Flowering plants such as Awlwort (*Subularia aquatica*), Bogbean (*Menyanthes trifoliata*), Shoreweed (*Littorella uniflora*), and

Bogbean (*Menyanthes trifoliata*) in the Dyffryn Ogwen.

Water Lobelia (*Lobelia dortmanna*) are typical of these acidic, low nutrient waters. Awlwort, Shoreweed and Water Lobelia are found in many lakes including Llyn Idwal and Llyn Ogwen, but Bogbean is more a plant of shallow ponds, especially in peat bogs. A less common species of acidic waters is Floating Water-plantain (*Luronium natans*) – now classed as a rare and vulnerable species, and restricted mainly to upland lakes such as Llyn Cwellyn and Llyn Tegid (Bala Lake). Other uncommon water plants in Snowdonia include the Six-stamened Waterwort (*Elatine hexandra*), found in Llyn Cwellyn, and the Least Bur-reed (*Sparganium natans*), which has recently been found among Water Horsetail (*Equisetum fluviatile*) in the waters of a flooded quarry near Marchlyn Reservoir.

In some of the less acidic and more nutrient enriched lakes such as Llyn Bychan and Llyn Goddionduon, water plants such as Alternate Water-milfoil (*Myriophyllum alterniflorum*), Broad-leaved Pondweed (*Potamogeton natans*), Common Club-rush (*Schoenoplectus lacustris*), Floating Bur-reed (*Sparganium angustifolium*), Common Spike-rush (*Eleocharis palustris*) and White Water-lily (*Nymphaea alba*) can be found.

BOGS AND FENS

In the wake of woodland clearance, and possibly linked with climate change around 5000-6000 years ago, extensive blanket bogs began to form in the wetter north and west of Britain. They are made of peat, a material formed when conditions are too wet and cold for plant remains to decay in the normal way. They are acidic due to the nutrient-poor status of rainwater and the almost continuous flow of water in these heavy rainfall areas washing out any soil nutrients. In some places the peat may be over five metres deep. Most of the blanket bog plant communities are characterised by *Sphagnum* mosses and a low canopy of heathland shrubs, mainly Common Heather and Cross-leaved Heath

(*Erica tetralix*), together with a few grasses and sedges. Purple Moor-grass (*Molinia caerulea*), Deergrass (*Trichophorum cespitosum*), and Hare's-tail Cotton-grass (*Eriophorum vaginatum*) are all likely to be present. At higher levels in the mountains, Crowberry (*Empetrum nigrum*) tends to replace Cross-leaved Heath, and in one or two places on the Berwyn, Cloudberry (*Rubus chamaemorus*) can be found at one of its most southerly British localities.

Hare's-tail Cotton-grass (*Eriophorum vaginatum*) in Cwm Idwal.

Many of these blanket bogs are still actively developing, but others have been modified. They are very susceptible to erosion, and when peatlands are drained for agriculture or forestry, the drying effect allows the normal processes of decomposition to resume. The practice of burning moorland vegetation – to create grazing for sheep or for grouse – can also damage the peat surface, and peat cutting for fuel or gardens can destroy them altogether. Since our present climate is no longer conducive to peat formation, except very locally, most of the damage is irreversible. The first signs of such change can be quite subtle, and detailed recording may be necessary to detect them. There may be, for example, a decline in bryophyte cover and an increase in some of the grasses, particularly Purple Moor-grass. This can eventually result in conversion of bog to wet heath and an impoverishment of the flora to the point that characteristic species are lost. The dominant heathland shrubs tend to be replaced by cotton grasses, leaving only a patchy cover of grasses, shrubs and bryophytes.

The best examples of blanket bog can be found on the large open spaces of the Migneint, the Berwyn and, on a smaller scale, in the Carneddau. Smaller examples also occur where flat areas coincide with impeded drainage, often in the bottoms of valleys such as Nant Ffrancon, or in cwms. In other cases they have developed in the basins of former lakes after a period of infilling – as at Cors Barfog near the Dyfi Estuary. Some of the less common species found on blanket bog include Few-flowered Sedge (*Carex pauciflora*), Tall Bog-sedge (*Carex magellanica*) and Lesser Twayblade (*Listera cordata*), all of which have been recorded on the Migneint. The poor soils and lack of nutrients on some wet sites mean that certain plants have to find nutrients, especially nitrogen, from alternative sources. Carnivorous or insectivorous plants are usually associated with tropical forests, but Snowdonia has its own miniature jungle in which several plants have evolved an ability to obtain at least some of their nutrients by capturing and digesting insects. The most abundant is the Round-leaved Sundew (*Drosera rotundifolia*): a small plant easily overlooked, but fascinating in its detailed structure. Each leaf is covered with hairs tipped by tiny sticky droplets, which ensnare insects and hold them for slow digestion by enzymes secreted at the centre of the leaf.

Round-leaved Sundew (*Drosera rotundifolia*) in Cwm Idwal.

The Common Butterwort (*Pinguicula vulgaris*) similarly traps and digests insects in its basal leaves, which extend like a rosette around the single flowering stalk. Both these species can be seen from the path around Cwm Idwal and in many other marshy places in Snowdonia. Two other insectiferous plants, Long-leaved Sundew (*Drosera intermedia*) and the Great Sundew (*D. anglica*) are far less common, but both can be seen at Cors Barfog.

In some bogs peat may have accumulated to such an extent that the surface is above the water table and plants must then obtain their water from rainfall rather than overland flow. These are known as raised bogs and although they can be a component part of blanket bogs, they are mainly restricted to lowland situations. The best example in

Common Butterwort (*Pinguicula vulgaris*) in Cwm Idwal.

Snowdonia is Cors Goch Trawsfynydd, which has suffered relatively little damage from drainage or cutting, and still supports the uncommon Bog Rosemary (*Andromeda polifolia*). Cors Arthog (Arthog Bog) near the mouth of the Mawddach Estuary is another example of this habitat, but with rather different vegetation. Much of it is dominated by Purple Moor-grass and Cross-leaved Heath, but the aromatic Bog Myrtle (*Myrica gale*) and White Beak-sedge (*Rhynchospora alba*) are common and there are several rarities including Bog Rosemary, Brown Beak-sedge (*Rhynchospora fusca*), Greater Spearwort (*Ranunculus lingua*) and Wavy St John's-wort (*Hypericum undulatum*).

In addition to blanket bogs and raised bogs, Snowdonia has some wonderful examples of flush bogs. These form wherever drainage systems such as lateral ground water movement becomes impeded, and are usually at the base of drainage slopes. Flush bogs are usually dominated by rushes such as Sharp-flowered Rush (*Juncus acutiflorus*) or Soft Rush (*J. effusus*), and some have a wide range of wetland species including Bog Stitchwort (*Stellaria alsine*), Bottle Sedge (*Carex rostrata*), Cuckoo Flower (*Cardamine pratensis*), Greater Bird's-foot-trefoil (*Lotus pedunculatus*), Marsh Bedstraw (*Galium palustre*), Starry Saxifrage (*Saxifraga stellaris*), Bog Pondweed (*Potamogeton polygonifolius*) and Marsh Violet (*Viola palustris*). Several *Sphagnum* species may also be present. It is possible to see flush bogs throughout Snowdonia, but there are particularly good examples on and around the Carneddau – in Cwm y Bedol and Cwm Llafar, for example.

Fens differ from bogs in being less acidic and in some cases they are distinctly calcareous. They are mainly found in lowland situations and there are some very good examples on Anglesey just north of Snowdonia. There are small fens in the uplands, however, where base-rich waters emerging from calcareous bedrocks run over impervious substrates and irrigate the peaty soils, giving rise to a distinctive community. Alongside the Miners' Track on Snowdon for example, small mires support Tawny Sedge (*Carex hostiana*), Dioecious Sedge (*Carex dioica*) and Flea Sedge (*C. pulicaris*) all characteristic of somewhat base-enriched conditions.

HEATHLANDS

Heathands are characterised by the presence of dwarf shrubs of the heather family. On dry heaths, species such as Common Heather (*Calluna vulgaris*), Bell Heather (*Erica cinerea*) and Crowberry (*Empetrum nigrum*) predominate, whereas on wet heaths Bell Heather is usually replaced by Cross-leaved Heath (*Erica tetralix*). As we have mentioned, blanket bogs may also support heathers, but they differ from heaths in that they occur on deep peat. Most heathlands are not entirely natural, but were created as a by-product of man's activities. It means that most heathlands below the natural tree line require human intervention to prevent reversion to scrub and eventually woodland. Burning, cutting and grazing are all possible methods of control. The large expanses of heather moorland on the Berwyn and many of the gentler mountain slopes in

Heathland at Cilgwyn near Pen y Groes.

Snowdonia are maintained either by grazing or by rotational burning. In Snowdonia, this is usually done for agricultural reasons, but it can also be for conservation, since heaths often provide habitat for a variety of interesting plant and animal species. Birds such as the hen harrier, black grouse, red grouse and merlin require large areas of heathland in which to breed and hunt. By burning on rotation the shrubs are prevented from becoming old and leggy, and space is created for younger, more vigorous plants.

Burning is not generally practised on the steeper slopes of mountains such as on the south-west face of Pen yr Ole Wen, where grazing is sufficient to prevent trees from becoming established – but not so heavy that the heather is replaced by grass. Not all heaths are man-made however, and there are areas of Snowdonia – the steep hillsides of the Rhinogau for example – where heath occurs naturally. Such areas are rather wet and only lightly grazed. Mountain Everlasting (*Antennaria dioica*) with distinctive white woolly hairs on the undersides of its leaves is one species that grows in such places. On these higher heathlands or moorlands the climate is too harsh for trees to grow, but because of the all pervading human factor, it is hard to know exactly where the transition between natural woodland and upland heath lies in today's climate, but it is probably somewhere in the range 500 to 700 m.

Other shrubs found on heaths, particularly dry heaths, include Bilberry (*Vaccinium myrtillus*) and Western Gorse (*Ulex gallii*), while on the Berwyn, as previously mentioned, it is possible to see the rare Cloudberry (*Rubus chamaemorus*) at its only Welsh locality. The plant resembles a low-growing Raspberry (*Rubus idaeus*) but produces orange coloured fruit when ripe. Western Gorse, as the name suggests, is more or less restricted to western heaths, and can provide a colourful spectacle when its bright yellow flowers form intricate mosaics with the purples of heather in August and September as seen, for example, on the southern slopes of Moel y Ci.

Some of the best wet heaths in Snowdonia can be found on the Carneddau and the Migneint. They generally contain mixtures of Cross-leaved Heath, Deergrass (*Trichophorum cespitosum*), heather and Purple Moor-grass, but other more spectacular species may be present. The Bog Asphodel (*Narthecium ossifragum*)

Bog Asphodel (*Narthecium ossifragum*) in marshes near Tregarth.

with its iris-like leaves, yellow flowers and orange anthers is one of the most beautiful, but the varying hues of blue-violet displayed by Heath Milkwort (*Polygala serpyllifolia*), Marsh Violet (*Viola palustris*), or the insectivorous Common Butterwort (*Pinguicula vulgaris*) can also be spectacular.

An uncommon species of this habitat is the alpine form of Juniper (*Juniperus communis* ssp. *nana*). More usually found in the Scottish Highlands, it occurs at a few sites in Snowdonia, including the slopes of the eastern ridge above Nantgwynant, to the south-west of the Snowdon summit, and on Y Lliwedd. This last site is by far the biggest, and contains a large portion of the total Welsh stock. At this altitude it occurs sometimes as isolated bushes, and at other times in 'colonies' alternately shaped and clipped by wind and grazing until some almost take on the shape of the rocks amongst which they grow.

Juniper (*Juniperus communis*) in Cwm Graianog.

GRASSLANDS

In the unenclosed uplands, heavy grazing often results in the replacement of dry heath by species-poor acid grassland, in which grasses such as Mat Grass (*Nardus stricta*), Sheep's Fescue (*Festuca ovina*), Common Bent (*Agrostis capillaris*) and Wavy Hair-grass (*Deschampsia flexuosa*) predominate. The poorest, most heavily grazed grasslands are often dominated by Mat Grass, which is avoided by sheep. They selectively graze the more palatable species, leaving extensive stands of the light coloured Mat Grass covering the hillside. However, even the poorest grasslands are usually enlivened by a scattering of the bright yellow flowers of Tormentil (*Potentilla erecta*) and the white flowers of Heath Bedstraw (*Galium saxatile*). Most of the flowering plants found in these acid grasslands are widespread, but in some of the high altitude examples, such as on Carnedd

Llewelyn, the rare Stiff Sedge (*Carex bigelowii*) can be found. Acid grasslands cover extensive tracts in the unenclosed uplands and in the higher ffridd zone, but are much more fragmentary at lower altitudes. In the enclosed lowlands, there are a few stands of species-rich acid grasslands, to the south of Llyn Trawsfynydd, for example, with species such as the uncommon Wood Bitter Vetch (*Vicia orobus*), Moonwort (*Botrychium lunaria*) and Mountain Pansy (*Viola lutea*). On more drought-prone soils, Sheep's Sorrel (*Rumex acetosella*) and a range of annuals, including Shepherd's Cress (*Teesdalia nudicaulis*) occur.

Unimproved dry grasslands on neutral soils are generally richer in species than those on acid soils. They may contain, for example, Crested Dogstail (*Cynosurus cristatus*), Red Fescue (*Festuca rubra*), Common Knapweed (*Centaurea nigra*), Bird's-foot-trefoil (*Lotus corniculatus*) and Red Clover (*Trifolium pratense*). They may be managed as pasture or closed from grazing in summer to produce a hay crop after the plants have set seed, and often provide an important refuge for species such as Greater Butterfly Orchid (*Platanthera chlorantha*) and the Adder's-tongue fern (*Ophioglossum vulgatum*), both of which have declined markedly as their habitat has been converted to more productive types of pasture. The best sites are widely scattered, from the Penmachno area in the north-east, to the flanks of Bwlch Mawr in the south-west, and they are usually small.

Much greater diversity can be found in calcareous grasslands or areas where drainage over such rocks results in flushing of the grassland, but such sites are uncommon in Snowdonia. The grasslands at the head of Llyn Idwal, and the slopes below Crib Goch and Crib y Ddysgl on Snowdon are of this type. Unfortunately, such grasslands are attractive to sheep, and often become impoverished by heavy grazing. They are not usually dominated by any one species, but rather have several present in moderate quantity. Typical species are Common Bent (*Agrostis capillaris*), Sweet Vernal Grass (*Anthoxanthum odoratum*), Sheep's Fescue (*Festuca ovina*), Autumn Hawkbit (*Leontodon autumnalis*), Harebell (*Campanula rotundifolia*), Self heal (*Prunella vulgaris*), Common Dog-violet (*Viola riviniana*), Wild Thyme (*Thymus polytrichus*), and Yarrow (*Achillea millefolium*). Some base-rich upland grasslands at high altitude, around Cwm Glas especially, include arctic-alpine species such as Alpine Bistort (*Persicaria vivipara*), Lady's-mantle (*Alchemilla glabra*), Alpine Meadow-rue (*Thalictrum alpinum*), Moss Campion (*Silene acaulis*) and Purple Saxifrage (*Saxifraga oppositifolia*).

Some rare and uncommon plant species also occur in grasslands, particularly where the grazing is not too intense. They are often in small numbers however, and long-term isolation can result in populations drifting apart genetically and in the long term possibly giving rise to new species. Although the emergence of a new species would normally take thousands of years, the process can be speeded up in species which reproduce mainly by self-fertilisation or hybridisation, processes which can produce new forms in a relatively short time. One such group of species is the eyebrights (*Euphrasia* spp.) and Snowdonia has its own endemic eyebright, *Euphrasia cambrica*, a tiny plant whose petals sometimes do not open to allow cross-pollination. The species is so difficult to find and positively identify that few survey data exist, but most records are from moderately base-rich, grazed grassland. The slopes to the north of Llyn Du'r Arddu may be a typical site. *Euphrasia rivularis*, another rare and slightly larger member of the group, is found only in North Wales and the English Lake District. It grows in damp, flushed grassland, streamsides and ledges, and can be found in the base-rich grasslands in Cwm Glas and on Moel Siabod. Both species are classified as rare in Britain and, as with other rarities, each has its own Conservation Action Plan.

In certain badly drained areas, such as the flood plains of the Afon Llugwy, tussocky stands of Purple Moor-grass can often be found. These wet grasslands are associated with bogs, but occur mainly in seasonally wet places rather than under permanent waterlogging. Most are on acid soils and include typical 'acid loving' species such as Common Cotton-grass (*Eriophorum angustifolium*), Petty Whin (*Genista anglica*), Heath Spotted-orchid (*Dactylorhiza maculata*) and Tormentil. Under appropriate management, these *Molinia* grasslands can support a varied flora and fauna. Of particular interest in Snowdonia are some of the more calcareous stands such as those on the banks of the Afon Conwy and Afon Eidda (south of Betws-y-Coed), the Afon Glaslyn (north of Tremadog), the Afon Prysor (east of Llyn Trawsfynydd) and at Tir Stent (near Dolgellau). Such sites are havens for a great variety of rather uncommon plants. Globe-flower, Marsh Hawk's-beard (*Crepis paludosa*), Melancholy Thistle (*Cirsium heterophyllum*), Petty Whin, Saw-wort (*Serratula tinctoria*) and Whorled Caraway (*Carum verticillatum*) may all occur. Orchids are represented by species such as Early Marsh-orchid (*Dactylorhiza incarnata*), Fragrant Orchid (*Gymnadenia conopsea*), Lesser Butterfly-orchid (*Platanthera bifolia*) and Northern Marsh-orchid (*Dactylorhiza purpurella*).

While many of these grassland species depend for their existence on grazing at moderate levels, the very heavy grazing that we have now is threatening their survival. Although sheep rearing has been the main farming activity in the Welsh uplands for a very long time, grazing intensity has increased greatly in the past 50 years or so, and is now the main factor limiting species distribution. Many species survive only because they have clung on at a few sites that are protected from grazing, either because of abnormally low stocking rates or because they are inaccessible to stock – on ledges or in crevices. The relationship between management and vegetation type can be seen in a number of places where fences make a sharp divide between heath on one side and short grassland on the other. The exclosures in Cwm Idwal and on Snowdon where sheep have been excluded since the 1950s show the effect very well.

WOODLANDS

Despite extensive losses, outstanding woodlands still survive in Snowdonia, with some of the best examples on steep valley sides, such as in Dyffryn Ffestiniog and Nantgwynant. They have usually survived because they were unsuitable for agriculture, or because the trees had intrinsic value, either for timber or as shelter for stock. Although most of them have been intensively managed in the past, it is likely that they are the descendents of the original wildwood. Species composition is dictated mainly by the underlying geology and hydrology, but the main tree species is usually Sessile Oak (*Quercus petraea*), although Downy Birch (*Betula pubescens*) is also a significant component of many woods. Ash (*Fraxinus excelsior*) may be dominant on some of the more nutrient-rich soils, such as in Coedydd Abergwynant, and Alder (*Alnus glutinosa*) may take over on wetter ground close to rivers or in water-logged hollows. A much less common type known as 'plateau alderwood' occurs on flushed hillsides and spring-lines such as at Coed Dolgarrog. Oak has always been the main timber tree, but Alder also has a history of use in the area, and was the basis of an industry making clogs for Lancashire mill workers. It was to revive this tradition that an Alder coppicing project began in Coedydd Aber in the mid 1990s. The main aim was to conserve the alderwood, since the act of coppicing rejuvenates the trees, and stimulates regeneration from seed. Unexpectedly though, the project developed into a small but sustainable industry for coppice timber used mainly for charcoal production.

In certain oakwoods – Coed Tremadog is an example – grazing has been low enough to allow an understorey of shrubs and small trees to develop. Hazel (*Corylus avellana*), Rowan (*Sorbus aucuparia*) and Holly (*Ilex aquifolium*) are common in such places, but at Coed Ganllwyd, the much rarer Alder Buckthorn (*Frangula alnus*) occurs. If any of the main canopy trees falls allowing light to penetrate to lower levels, birch can become temporarily dominant because of its

Woodland in Nantgwynant.

speed of growth but, given time, oak will usually reassert itself. In Alder woods various species of willow or Downy Birch may become established, and in some places such as Hafod Garregog, Nant Ffrancon and around Llyn Cororion, Downy Birch is the dominant species.

The composition of the ground flora in woodlands depends on factors such as soil fertility, drainage, grazing intensity and shading. However, with our humid climate, many of Snowdonia's oakwoods are dominated by luxuriant growths of mosses, liverworts and ferns, with flowering plants playing a subordinate role.

Wood-sorrel (*Oxalis acetosella*) in Bryn Meurig.

The bright yellow flowers of Common Cow-wheat (*Melampyrum pratense*) and the white flowers of Wood-sorrel (*Oxalis acetosella*) may be encountered, while in some of the less rocky oakwoods on acid soils, such Coed Camlyn and Coed Dinorwic, Heather (*Calluna vulgaris*) and Bilberry (*Vaccinium myrtillus*) may be abundant. The ground flora of these woodland heaths can be heavily modified by grazing, resulting in heathy shrubs sometimes being replaced by grasses, particularly Wavy Hair-grass (*Deschampsia flexuosa*). In the more rocky, bryophyte-rich woodlands, such as those around Capel Curig, the rare Alpine Enchanter's-nightshade (*Circaea alpina*) may be seen, but it is easily confused with Intermediate Enchanter's-nightshade (*C. x intermedia*), a hybrid between Alpine and Enchanter's-nightshade (*C. lutetiana*). The hybrid is quite a vigorous plant, which can spread through vegetative reproduction, and it has been recorded in several places including Coed yr Allt Goch, north of Betws-y-Coed and around the shore of Llyn Tegid (Bala Lake). Alder woods are only slightly better off for flowering plants, but may contain Meadowsweet (*Filipendula ulmaria*), Opposite-leaved Golden Saxifrage (*Chrysosplenium oppositifolium*) and Yellow Pimpernel (*Lysimachia nemorum*). In southern Snowdonia the rare Touch-me-not Balsam (*Impatiens noli-tangere*) can be found in Alder and other wet woodlands.

It is only in some of the less acidic woodlands such as Coed Gorswen, Coed Dolgarrog and Coed Tremadoc that we see more species diversity in terms of trees and ground flora. The trees may include Ash, Sycamore (*Acer pseudoplatanus*), or rarely Small-leaved Lime (*Tilia cordata*). English Elm (*Ulmus procera*) also occurs in these woods, but since the ravages of Dutch Elm disease, it no longer makes a contribution as a mature tree – most specimens being killed by the disease before they reach maturity. It maintains a presence, though, in the form of saplings, which continue to arise as suckers from trees that are dead above ground, but still have live roots. The Wych Elm (*Ulmus glabra*) is not immune to the disease, but seems less susceptible, and can still be found occasionally in these woods. The large stands of Beech (*Fagus sylvatica*) in woodlands such as at Coed Dolgarrog or the woods above the A498 south of Aberglaslyn are not native to Snowdonia, but they are naturalised and in some areas even dominate the native woodlands.

The ground flora of these base-rich woodlands is richer than in most of the upland oakwoods on acid soils, and may include species such as Broad-leaved Helleborine (*Epipactis helleborine*), Early-purple Orchid (*Orchis mascula*), Dog's Mercury (*Mercurialis perennis*), Enchanter's-nightshade, and Ramsons

Globe-flower (*Trollius europaeus*) in Cwm Idwal.

(*Allium ursinum*) which, on damp spring evenings, fills the air with the pungent aroma of garlic. Less common species can also be found, such as Moschatel (*Adoxa moschatellina*), Woodruff (*Galium odoratum*), Wood Millet (*Milium effusum*) and Sanicle (*Sanicula europaea*). In Coed y Gofer it is also possible to see the very rare Narrow-leaved Helleborine (*Cephalanthera longifolia*). Perhaps surprisingly, the Globe-flower (*Trollius europaeus*) is also associated with base-rich woodland, and has been recorded in Coed Gors y Gedol.

The most stunning sight in these woodlands must be the blue carpets of springtime Bluebells (*Hyacinthoides non-scripta*), opening beneath the still leafless trees. Few people realise that Britain contains a large proportion of the world's population of this species, and that we have an important role in its conservation. Bluebells have always been a favourite among flower pickers, but the more recent practice of stripping bulbs from woodlands has made it necessary to institute statutory protection under the Wildlife and Countryside Act.

ARTIFICIAL HABITATS

Much of what is often regarded as derelict land – the old slate quarries and the spoil heaps from abandoned heavy metal mines – can be of considerable interest for plants. The main ores contained copper, lead or zinc, and the waste tips produced by their extraction are widespread. In high concentrations these metals are toxic to most plants, and tips containing them are usually colonised only by a few plants that have evolved a special tolerance. Some, like Red Fescue (*Festuca rubra*), are ecotypes of common species but others are specialists, and grow only where these elements are present. The lead/zinc waste tips just north of Betws-y-Coed for example, are almost the only sites in Snowdonia for the rare Alpine Penny-cress (*Thlaspi caerulescens*). Outside Snowdonia, it is restricted mainly to eastern counties.

The many old slate quarries of Snowdonia also support their share of uncommon plants. Slate tips, for example, are a habitat almost identical to natural scree slopes and, although mainly important for ferns and bryophytes, flowering plants such as Bilberry and Foxglove (*Digitalis purpurea*) may also be present. However, a number of uncommon species, including Bog Asphodel (*Narthecium ossifragum*), Least Bur-reed (*Sparganium natans*), Star Sedge (*Carex echinata*) and White Beak-sedge, have been recorded in flooded quarries. Fewer species may be present in the drier quarries, especially where the soils are thin, but the uncommon Small Cudweed (*Filago minima*) has been found at such a site near Marchlyn Bach.

CONSERVATION

As the most southerly high mountain block in Britain, Snowdonia is an important area for conservation, and has a number of small populations of species that normally grow at higher latitudes or altitudes. They and some of the vegetation communities in which they occur are therefore at the edge of their range, and can provide valuable information on the probable effects of possible future climate change, such as global warming. It is these peripheral communities, that will show the first effects and so indicate how the more central areas will eventually react.

Despite their extent, the grasslands we see over most of the mountains in Snowdonia are not natural, but the product of centuries of grazing by sheep, cattle and goats. Grazing intensity has increased in recent years, particularly since the Second World War when farmers were induced to increase their stocking rates through a system of subsidies. The many years of heavy grazing has prevented natural regeneration of woodland, and destroyed much of the heathland, leaving only fenced areas and the more inaccessible slopes and cliffs able to support 'natural' vegetation communities. A major conservation aim in Snowdonia therefore, is to reduce the effects of grazing and increase the spread and health of such vegetation.

Not only have domestic stock increased, but so have people, with more and more visiting Snowdonia every year. If the trend continues, there is a real danger of their degrading the very environment that they find so attractive. The era of collecting has largely gone, but indirect damage can occur due to erosion from thousands of walkers' boots, the construction of road and rail links, and the sheer disturbance that such numbers can cause to wildlife.

Atmospheric pollution is also of major concern in many parts of Wales including Snowdonia. Emissions of oxides of sulphur, the main contributor to acid rain, have been responsible for the loss of many lichen species close to emission sources, and for wider effects on freshwater ecosystems. Such pollution has lessened in the last 20 years, but long-range transport and the high rainfall in Snowdonia mean that the amounts in Snowdonia are still high. Over this same period, releases of oxides of nitrogen were increasing largely due to increases in road traffic. Nitrogen contributes to acid rain and acts almost as an unwanted fertiliser on sensitive vegetation communities such as heaths and bogs, resulting in the invasion of competitive plant species. Research is therefore being undertaken into the impact of the combined effects of nitrogen deposition and sheep grazing on this vegetation. Other land uses, such as the planting of coniferous trees, can exacerbate acidification due to the fact that trees tend to 'scavenge' pollutants from the atmosphere. The effect is so serious in some parts of Wales that attempts are being made to counter it by adding powdered limestone to lake catchments. However, trials in which lime was applied to wet flush areas demonstrated that bog mosses and some other groups of plants can be damaged as much by the lime as by the acidification.

Other threats arise from introduced and invasive species and, in Snowdonia,

Rhododendron (*Rhododendron ponticum*) is by far the most notorious example. A native of southern Europe and south-west Asia, it escaped from cultivation many years ago, and is now well established in several parts of Snowdonia, but especially on the hillsides around Beddgelert and Maentwrog. Although it is beautiful in spring, when the Aberglaslyn Pass becomes a riot of pink and purple blossom, the lack of a natural check to its growth, its unpalatability to stock, and its ability to shade out other species, have all led to its being regarded as a serious pest. Were it not for continuous efforts at control by various conservation bodies, Rhododendron would by now cover most of the valley sides in Snowdonia, gradually taking over what should be heather covered hillsides or native oakwood. Control is made difficult though by two factors. First, it is difficult to kill with a foliar spray because the leaves are protected by a thick cuticle, and second, the cut stumps have the power to grow new shoots. A method has been devised which is effective, but it entails a great deal of work, with repeat visits to prevent regrowth.

Another introduced, but less conspicuous species is the New Zealand Willowherb (*Epilobium brunnescens*). First collected in 1908, it has spread to all kinds of damp, bare ground in the British Isles and has made a home in the rocky slopes of Snowdonia. It is still spreading, and its small pink flowers can be seen on any walk through rocky areas, but unlike Rhododendron, it does not appear to be threatening any native plants.

More worrying introductions include Oxford Ragwort (*Senecio squalidus*) and Japanese Knotweed (*Fallopia japonica*). The former was first recorded in 1794. After spreading along railway lines from the Oxford botanic garden where it was introduced, it has now become widespread in England and Wales. In Snowdonia, it is well established in some semi-improved fields. Japanese Knotweed, is a more recent introduction, first found growing wild in Britain in 1886. It has now spread to the point that it poses almost as large a problem as Rhododendron, but mainly along roadsides and watercourses rather than on hillsides. Control is again problematical, with even small sections of stem having the ability to root and form a new plant. Until recently this was its only means of dispersal as no viable seed was being produced, but it seems that this may soon change. A hybrid plant with the ability to set seed has been discovered near Dolgellau and, if it has retained the vigour of Japanese Knotweed, we could be facing a major problem.

The conservation of the flora of Snowdonia is tied up with general habitat conservation and cannot be entirely divorced from it. As a general rule in conservation, if the habitat is in good condition, then so also will be the species it supports. There is still so much we do not know about some of these species, however, even their basic biology and ecology, and so for now we can only try to ensure that we have viable populations which can reproduce effectively. There is increasing pressure to translocate threatened plants or reintroduce populations to sites where they once grew. However, too much intervention becomes a little like gardening, and can obscure the natural distribution of plants. If we can increase the size of the populations of these threatened and vulnerable species by habitat management, then the need for intervention may well diminish.

There are a number of flowering plants in Snowdonia which are protected by law under the Wildlife and Countryside Act, 1981 (as amended). They include:

Dwarf Spike-rush	*Eleocharis parvula*
Slender Cotton-grass	*Eriophorum gracile*
Bluebell	*Hyacinthoides non-scripta*
Welsh Mudwort	*Limosella australis*
Snowdon Lily	*Lloydia serotina*
Floating Water-plantain	*Luronium natans*
Tufted Saxifrage	*Saxifraga cespitosa*
Spiked Speedwell	*Veronica spicata*

Protected species that appear to have been lost from Snowdonia include:

Field Wormwood	*Artemisia campestris*
Deptford Pink	*Dianthus armeria*
Field Eryngo	*Eryngium campestre*
Pennyroyal	*Mentha pulegium*

ADDITIONAL READING

Condry, W. 1966. *The Snowdonia National Park*. The New Naturalist. Collins, London.

Evans, P. 1932. Cader Idris: a study of certain plant communities in south-west Merionethshire. *Journal of Ecology*, 20: 1-52.

Godwin, H. 1956. *History of the British Flora*. Cambridge University Press.

Ratcliffe, D. A. 1959. The vegetation of the Carneddau, North Wales. *Journal of Ecology*, 47:371-413.

Ratcliffe, D. A. & Thompson, D. B. A. 1988. The British Uplands: Their ecological character and international significance. In M. Usher & D. B. A. Thompson (eds.), *Ecological Change in the Uplands*, Oxford: Blackwell Science.

Raven, J. & Walters, M. 1956. *Mountain Flowers*. The New Naturalist. Collins, London.

Stewart, A., Pearman, D. A. & Preston, C. D. 1994. *Scarce Plants in Britain*. Joint Nature Conservation Committee. Peterborough.

Wilson, A. 1927-28. The Botany of Tal-y-Fan. *Proceedings of the Llandudno and Colwyn Bay Field Club*, 1927-28.

Woodhead, N. 1927-28. The Botany of Llyn Ogwen and Llyn Idwal. *Proceedings of the Llandudno and Colwyn Bay Field Club*, 1927-28.

FERNS

DEWI JONES

INTRODUCTION

Ferns constitute a division of flowerless, vascular plants. They have a fossil record stretching back over 350 million years to the upper Devonian Period, and became a major component of the Earth's vegetation during the Carboniferous Period, some 300 million years ago. Ferns or fern-like plants, such as the Devonian fossil *Protopteridium*, were among the first plants to develop leaves, and in the case of *Protopteridium* they were probably nothing more than flattened branches, but they were the first step in the evolution of the elaborate leaves that we see in today's ferns. Leaves have also evolved in other primitive plants, such as the clubmosses, but those of ferns are unique among seedless plants in having a complex vascular system not unlike that of flowering plants. Nevertheless, certain living species seem remarkably similar to their Palaeozoic ancestors. The Royal Fern (*Osmunda regalis*) for example, is almost indistinguishable from ferns that were living some 250 million years ago. It is not surprising that some of these species are described as living fossils.

Like other lower plants, the ferns have a complex life cycle involving two distinct generations. One, the sporophyte, is a large and long lived plant that to most people *is* a fern, and the other, the gametophyte – also called the prothallus – is a tiny green heart-shaped structure. The sporophyte gives rise to spores, which in the more typical ferns such as the Male-fern (*Dryopteris filix-mas*), are produced in structures known as sori on the under surface of mature fronds. Germination of a spore leads to the production of a prothallus, which is the sexual phase responsible for the production of male and female organs – the antheridia and the archegonia. The fern prothallus differs from its clubmoss equivalent in the fact that it contains chlorophyll and, under the right conditions, it can sustain itself for long periods – perhaps even years. Sexual union occurs when sperm produced in the antheridia swim through a water film to fertilize egg cells in the archegonia, and development of a new fern plant then follows. In some species, gametophytes are produced in greater numbers than sporophytes, to the extent that at some sites, a species may exist only as a gametophyte. Although the Killarney Fern (*Trichomanes speciosum*) is very rare in Snowdonia, it seems that the gametophyte is present at far more sites than the sporophyte.

THE HISTORY OF FERN EXPLORATION AND THE IMPACT OF THE VICTORIAN FERN CRAZE

Published records for ferns in Snowdonia start to appear in the 17th century. Edward Lhuyd (of Snowdon Lily fame) visited Snowdonia in the late 17th century and made records of various species that were later published by the celebrated naturalist John Ray in his book *Synopsis Methodica Stirpium Britannicarum* (1690-1724). In comparison with other plant groups, few of the early British botanists took any particular interest in ferns, but during the 1830s field botany became fashionable among members of the British upper and middle classes, and this coincided with a major trend which came to be known as the Victorian fern craze. But botany was not just a rich man's hobby; even as early as the late 18th century, there were working-class groups of self-taught naturalists who had established botanical societies within their districts holding field excursions and indoor meetings where plant specimens were discussed and identified. However, apart from James Bolton's finely illustrated *Filices Britannicae* (1785-90) there were no books dealing exclusively with ferns until George William Francis published his *Analysis of British ferns and their allies* in 1837. Being reasonably priced, this book appealed to a wide range of fern enthusiasts and, with its useful text and meticulously drawn copperplate engravings, it became so popular that it ran into three further editions. During the autumn of the same year, Edward Newman went on a walking tour in the Welsh mountains and decided to learn more about the ferns he encountered. He gathered several hundred fronds and, in a quiet inn at the end of the day with no taxonomic guide to assist him, set about arranging them into portions of supposed species. After carefully transporting them home he states that he 'planted them with care, for the purpose of obtaining a more correct knowledge of the variations to which they were subject'. Newman was later to produce a large work entitled *A History of British Ferns*, in which he described each species in detail and, in some cases, gave advice on cultivation. Line drawings added to the appeal of the work. He also described his favourite hunting grounds – in words and in some cases with attractive sketches. His description of Cwm Idwal clearly demonstrates his passion for the site.

> I have more than once mentioned Cwm Idwell [sic] as a station for ferns: below is a very humble attempt to give an idea of this wild spot. It was sketched in a memorandum-book, and carried on my back among fern-fronds for many a weary mile. Cwm Idwell [sic] is a vast semi-circular rampart of rock, near the middle of which, invisible at a distance, is the perpendicular fissure called Twll dhu [sic]: through this falls a mountain stream, which, emerging at the foot, wanders, amongst fragments of disrupted rock into Llyn Idwell, [sic]- that dark, still lake which reposes in the natural basin; issuing thence, it joins the waste water of Llyn Ogwen, and the united stream flows though the mighty pass of Nant Frangon [sic] to the sea. In Llyn Idwell [sic] grow Isoetes [Quillwort], Subularia [Awlwort] and Lobelia [Water Lobelia] on the broken ground about the lake, Lycopodium alpinum [Alpine Clubmoss], L. selago [Fir Clubmoss], L. selaginoides [Lesser Clubmoss] and L. clavatum [Stagshorn Clubmoss] every conceivable form of

Cystopteris fragilis [Brittle Bladder-fern], with Allosorus crispus [Parsley Fern] and Hymenophyllum unilaterale [Wilson's Filmy-fern] a little higher up, Polystichum lonchitis [Holly-fern], Asplenium viride [Green Spleenwort], Rhodiola rosea [Roseroot] and alpine Thalictrum [probably Alpine Meadow-rue] and, that rarity of rarities, Anthericum serotinum [Snowdon Lily] and, still higher, above and beyond the summit that we see, Woodsia Ilvensis and Lycopodium annotinum [Oblong Woodsia and Interrupted Clubmoss]. Oh! It is a matchless place for a botanical ramble!

Unfortunately, during the 19th century the systematic stripping of plants from their natural habitats became common practice, heralding the start of the Victorian collecting mania in which ferns were favoured above all other plant groups.

In Snowdonia, as in other parts of the British Isles, botanists had been regular visitors since the first half of the 17th century, but by the 19th century they found themselves having to compete for their specimens with plant collectors. As demand increased, so did the nurseries which specialised in growing ferns, and they would replenish their stocks with wild plants. Local people also started collecting ferns, among them the mountain guides, copper-miners, slate quarrymen and women from the working classes, who found that there was money to be made in selling plants to visitors.

The Holly-fern (*Polystichum lonchitis*) was undoubtedly *the* souvenir fern of Snowdonia, and by the mid 19th century, guides were selling specimens from huts on the summit of Snowdon at sixpence a root. Although the effect on the wild populations of these plants must have been severe, the Holly-fern survived in a few inaccessible places, and still grows today in gullies, beneath rocks and on sheltered base-rich ledges. It was seen by Edward Lhuyd during his late 17th-century visits to Snowdonia, and his record from 'Clogwin y Garnedh y Crib Goch Trigfylchau' appeared in John Ray's *Synopsis Methodica Stirpium Britannicarum*, and in *Camden's Britannia*. According to John Ray's *Synopsis*, Lhuyd also discovered the Forked Spleenwort (*Asplenium septentrionale*)

Forked Spleenwort (*Asplenium septentrionale*) in Coedmawr west of Llyn y Parc.

growing on the top of Carnedd Llewelyn. Ray mentions finding it in the Llanrwst area, but his description of the site 'In muris antiquis *Llan-Dethylae* uno circiter milliari a *Llan-Rhoost* aquilonem versus' is confusing. No place by the name Llan-Dethylae is known – in the Llanrwst area or, indeed, anywhere in Snowdonia. The species favours hard lime-free rocks or, in a few instances, the interstices of dry-stone walls, but it also occurs abundantly on old lead-mine waste just west of Llanrwst, and in the brickwork of nearby ruined buildings. These sites are at low altitudes, a fact that prompted Raven & Walters (1965) to claim that Forked Spleenwort is not a true mountain plant in Snowdonia. However, Edward Lhuyd recorded the species 'On top of Carnedh Llewelyn, near Llan Lhechyd [sic]' (Camden, 1695) and, during the 19th century, bryologist William Wilson of Warrington recorded it from 'near Twll Du' at an altitude between 600 m and 700 m. The present author found it at an altitude of 570 m on the Moel Hebog range, confirming that, in Snowdonia at least, it really is a mountain plant.

Two other mountain ferns, Alpine Woodsia (*Woodsia alpina*) and Oblong Woodsia (*Woodsia ilvensis*), were also much sought after by the early fern collectors, but both are rare and elusive. In the past, visiting botanists and herbarium collectors tried to save time by employing local people either to guide them to the localities, or to gather specimens on their behalf. For example, the London botanist William Bennett sought the help of local people when he visited the Snowdon district in 1849. He spent the best part of a day searching Clogwyn Du'r Arddu with one of the younger guides and, after gathering some of the commoner fern species, decided to look for rarities *en route* to Snowdon summit. As they stood near Bwlch Glas, overlooking Cwm Glaslyn and the precipices of Clogwyn y Garnedd, they met the veteran plant-hunter William Williams of the Royal Victoria Hotel, Llanberis. Bennett enquired of Williams the site of the Alpine Woodsia, and the guide obligingly pointed out the 'exact spot' far below amid massed cliffs and boulders, knowing quite well that the stranger would never find the place. As Bennett began the descent, Williams promised to post him a specimen should he fail and kept his word. The packets that he used for the purpose are still preserved in the British Museum (Natural History). Bennett did manage to locate the Holly-fern on that occasion, only to find when he reached the summit, that the guides there had some for sale, but they knew nothing about the Woodsia. Such incidents would have been common in Snowdonia at the time.

Undoubtedly, the rarest and most coveted of all the ferns of Snowdonia was and still is the Killarney Fern. It is a species of dark damp caves and shaded waterfalls, and even today retains an almost mythical quality. In 1863 J. F. Rowbotham of Manchester claimed to have found a large number of well-developed plants growing in a very moist cave, 'from whose top and sides numerous fronds hung in the manner of a silky veil'. This was the first recorded Killarney Fern in Snowdonia, the whereabouts of which is still shrouded in mystery as the locality has never been rediscovered. One of the specimens gathered from it measured 18 inches in overall length (about 45 cm), suggesting that it was a well-established colony.

In 1887, John Lloyd Williams, then a schoolmaster at Garn Dolbenmaen School, published a note in the *Journal of Botany* recording his discovery of a Killarney Fern colony on Moel Hebog near Beddgelert and, despite being stripped bare of fronds by an unknown fern-hunter shortly after its discovery, this colony has survived to the present day.

Without doubt, the most famous of the early botanical explorers in Snowdonia was Edward Lhuyd – the man who discovered the Snowdon Lily. Perhaps the first reference to his explorations occurs in a letter dated 1682 in which he mentions a visit he made to Cwm Idwal. His records on that occasion were listed in *Camden's Britannia*, but this was not his only pioneering work, and he went on to make new finds at several other key sites including Clogwyn Du'r Arddu, Cwm Glas Mawr and Clogwyn y Garnedd. Some outlying areas, though, remained unexplored for many years after Lhuyd's time, and it was not until the late 19th century that coverage extended to the Cwm Llefrith / Moel yr Ogof area – a site now recognised as one the richest in Britain for fern species. The pioneer here was John Lloyd Williams who, after a varied career, became a leading algologist and then Professor of Botany at Aberystwyth University in 1915.

Closely related to the Killarney Fern are the two filmy ferns, Tunbridge Filmy-fern (*Hymenophyllum tunbrigense*) and Wilson's Filmy-fern (*H. wilsonii*). These are our smallest native ferns, and in general appearance resemble a moss or leafy liverwort. Both grow in wet shady habitats, but the former species is much the less common. They are rarely found in upland situations, although there is one place on the Moel Hebog range where the two species have been found growing together. The distinction between them was initially unrecognised, and it was

Tunbridge Filmy-fern (*Hymenophyllum tunbrigense*) in Ceunant Mawr near Llanberis.

only in 1829, when William Wilson saw them growing together in Killarney, that he realized that there really were two distinct forms. The name Wilson's Filmy-fern was assigned to the commoner of the two, while the more elegant Tunbridge Filmy-fern retained the name that until then had been used for both species.

The Broad Buckler-fern (*Dryopteris dilitata*) is common in shady places in woodlands throughout Snowdonia, but a related and much scarcer species, the Northern Buckler-fern (*D. expansa*), also occurs. It was first recorded in Britain by Thomas Moor in his book *A Popular History of British Ferns* (1859), but only as a variety of *D. dilitata*. His record was from Ben Lawers in Scotland, but a record for Snowdonia subsequently appeared in J. E. Griffith's *Flora of Anglesey and Carnarvonshire* (1894/5), and it has now been found in several places including the west cliff of Carnedd Dafydd, Cwm Cywion (on the Glyderau) and Cwm Glas (on Snowdon). Griffith was a local amateur botanist who made his living as a chemist in Bangor before a change in circumstances allowed him to retire early and devote himself to his other interests including archeology, antiquities and photography as well as botany.

In the lowlands, the Marsh Fern (*Thelypteris palustris*) has been recorded in at least one marshy area in Snowdonia. Arthur Aikin recorded it 'in a moist dell near the village of Llanberis' on his travels through Snowdonia in 1796, and his record appears in William Withering's *Arrangement of British Plants* (1796). Later William Bingley (1798 and 1801) also apparently found it at the same site, and on the strength of Bingley's statement, J. E. Griffith quoted the record in his 1894/5 flora despite not having seen it there himself. Bingley's description, however, was worded exactly like Aikin's, throwing some doubt on to whether he also ever personally observed the plant at this site. At any rate, one must assume that by Griffith's time the plant had probably disappeared from the Llanberis site, possibly through land drainage, and there have been no further records for Snowdonia.

The Royal Fern, also known as Water Fern or Flowering Fern, is also a species of lower altitudes. John Ray (1670) writes that it was 'observed in boggy places . . . in Wales'. Unfortunately, it is too attractive for its own good, and wild populations have been badly depleted by collecting for garden use – a practice that has sadly continued even up to quite recent times. It has consequently been lost from many of its former sites, and is now mainly confined to bogs on protected sites such as Cors Ty'n y Mynydd, Cors Gyfelog and Cors Graianog, and to streamsides in steep wooded gorges.

Like the Royal Fern, the Hart's-tongue (*Asplenium scolopendrium*) is not usually regarded as a mountain species, and although it does occur, it is limited to very sheltered sites with at least some lime in the substrate. However, William Salesbury (c.1520-91) records his findings of some excellent specimens in his manuscript herbal. The following passage translated from Welsh constitutes the first Welsh record.

> The finest I ever saw grew either side of 'bwll y vwyall' in a wooded glen on land belonging to Tudor ap Robert . . . There I gathered fronds up to half a yard within

the breadth of a hand some with forked ends and some thrice forked into four ends similar to a stag's horn.

John Lloyd Williams found this fern on Moel Hebog, and although it was not an especially dramatic discovery, it had an interesting sequel in that it was the event that inspired him to explore further afield, and eventually to discover the Killarney Fern.

The Victorian fern craze, beginning in the 1830s, was a purely British phenomenon, and the early fern books were mostly written with this obsession in mind. They provided not only taxonomic descriptions to help in species identification, but also gave instructions on methods of collecting, drying, pressing and mounting of specimens, with added tips on cultivation. The obsessive collecting of the period resulted in the near-extermination of some species, and caused great concern among local botanists. They made it clear that they were disgusted at the attitude of some of the herbarium collectors in which rare species were being deliberately targeted. J. E. Griffith commented in his *Flora of Anglesey and Carnarvonshire* (1894/5) that the Holly-fern 'is disappearing fast, and I am afraid it will soon be extinct.' J. L. Williams, on the other hand, admitted in his 1945 autobiography that he was, for a short period, involved in collecting for herbaria, and recalled an incident which convinced him that he should mend his ways. He was visited by a clergyman who had just published a flora of one of the English counties and was anxious to secure a specimen of a certain plant for his herbarium. Williams accordingly led him to the appropriate site, and then offered to show him another nearby site where the uncommon Forked Spleenwort grew. The collector's reply – that he already had a specimen – made Williams realize that the man's only interest in the species was as a possible addition to his herbarium.

IMPORTANT HABITATS FOR FERNS IN SNOWDONIA

Snowdonia, with its remote *cymoedd* (corries), rugged screes, steep cliffs, lakes, moorlands, heathlands, grasslands, woodlands, and its great variety of rock types, has one of the most varied and interesting fern floras in Britain. The region also has significance as the southernmost British locality for several rare mountain species, including Holly-fern, Alpine Woodsia and Oblong Woodsia. The following paragraphs deal with some of the more important habitats.

MOUNTAIN TOP FERNS

Very few ferns can live on mountain tops, but it is a curious fact that within Snowdonia, the species that do occur in such places are not strict alpines. The Parsley Fern (*Cryptogramma crispa*), for example, grows close to the summit of Carnedd Llewelyn and among the rocks of Snowdon's summit cairn. Also in this zone one may find Lady-fern (*Athyrium filix-femina*), Hard-fern

Parsley Fern (*Cryptogramma crispa*) in Cwm Idwal.

(*Blechnum spicant*), Beech Fern (*Phegopteris connectilis*) and Broad Buckler-fern, although they grow well only where there is some shelter – typically under loose rocks. The true alpines generally occupy the lower crags where the rocks are to some degree base-rich, and there is some shelter from the wind.

Beech Fern (*Phegopteris connectilis*) in Dyffryn Ogwen.

Green
Spleenwort
(*Asplenium
trichomanes-
ramosum*) in
Cwm Idwal.

Holly-fern
(*Polystichum
lonchitis*) in
Cwm Glas
Mawr.

FERNS OF MOUNTAIN CLIFFS AND LEDGES

The ferns found on cliffs and ledges depend very much on the rock type. Where there are only acidic rocks, the fern flora is limited to species such as the Mountain Male-fern (*Dryopteris oreades*), the Parsley Fern, and – rather rarely – the delicate Maidenhair Spleenwort (*Asplenium trichomanes* subsp. *trichomanes*). Many of the base-rich sites in contrast are famous for their species diversity and for the fact that they hold some extreme rarities. Sites such as Cwm Idwal, Cwm Glas Mawr and Clogwyn Du'r Arddu are regarded as classic in this respect. The ferns of these places are often mixtures of species that normally occupy other habitats. There are the true arctic-alpines, including Green Spleenwort (*Asplenium trichomanes-ramosum*), Brittle Bladder-fern (*Cystopteris fragilis*), Northern Buckler-fern, Mountain Male-fern, Holly-fern, Moonwort (*Botrychium lunaria*), Alpine Woodsia and Oblong Woodsia.

Alpine Woodsia (*Woodsia alpina*) in Cwm Glas Bach.

Oblong Woodsia (*Woodsia ilvensis*) on Moel yr Ogof.

Then there is a group of species that are thought to be woodland relicts, typical of wooded ravines in the uplands, and which require shade, moisture and high humidity. Among them are Lady-fern, Hard-fern, Common Scaly Male-fern (*Dryopteris affinis*), Broad Buckler-fern, Male-fern, Oak Fern (*Gymnocarpium dryopteris*), Beech Fern, Intermediate Polypody (*Polypodium interjectum*),

Common Polypody (*Polypodium vulgare*) and Hard Shield-fern (*Polystichum aculeatum*). It may be of interest to note that a certain growth form of the latter, known as *Polystichum aculeatum* forma *cambrica*, is superficially similar to Holly-fern and has been mistaken for it.

Oak Fern (*Gymnocarpium dryopteris*) in Ceunant Mawr near Llyn Gwynant.

Polypody (*Polypodium* sp.) in Coedydd Aber.

FERNS OF BOULDER FIELDS, CWMS AND GORGES

Some of the most unusual landscapes occur where boulders have fallen away from an eroding cliff and built up into fields of boulders on the slopes below. Perhaps the best examples in Snowdonia are in Nant y Benglog, especially under the cliffs of Gallt yr Ogof. Such places harbour a variety of common fern species including Lady-fern, Common Scaly Male-fern, Common Polypody and Intermediate Polypody, and slightly less common ones like the Lemon-scented Fern (*Oreopteris limbosperma*) – growing on the steeper slopes close to mountain streams. The rare Northern Buckler-fern may also occur in pockets of soil with heather and Hard Fern among large blocks of granite. Although it is usually associated with the higher mountains including places like Cwm Glas Mawr, Cwm Cywion and Cwm Pen Llafar, it can also grow at lower altitude as on Yr Eifl and other hills of the Lleyn Peninsula.

Lemon-scented Fern (*Oreopteris limbosperma*) in Ceunant Mawr near Llyn Gwynant.

Despite their names, neither Beech Fern, nor Oak Fern is restricted to beech or oak woodlands, and both can be found nestling among rocks in more open habitats, especially where there is a combination of shelter and low acidity. Both species occur, for example, in Cwm Idwal and Ceunant Mawr (north of Llyn Gwynant). The Hay-scented Buckler-fern (*Dryopteris aemula*) with the fragrance of new-mown hay can also be found in these situations, as well as in some of the boulder fields on the flanks of Moel Hebog, but it is far less common than either the Beech Fern or the Oak Fern. Also rare on stony slopes is the Limestone Fern

(*Gymnocarpium robertianum*), a species with an even stronger requirement for calcareous conditions than its close relative the Oak Fern. According to *The Botanist's Guide* (1805) published by D. Turner and L. W. Dillwyn, it was once found in Cwm Idwal by J. W. Griffith, but the herbarium specimen is now thought to have been labelled with the wrong place name. The Rigid Buckler-fern (*Dryopteris submontana*), with similar habitat requirements to the Limestone Fern, was recorded in J. E. Griffith's flora from both Cwm Idwal and Cwm Glas Mawr, but has not been seen in either place in recent years.

FERNS OF UPLAND SCREE

Despite a barren appearance, scree actually offers many niches for plants, especially ferns. Where scree is particularly mobile however, few ferns have the chance to grow to maturity, and only juvenile specimens may be present. Coupled with the fact that the various species may be hard to distinguish as juveniles, it creates something of a taxonomic problem. Lady-fern and buckler ferns often look similar to the Brittle Bladder-fern, which in turn can be mistaken for a woodsia. However, species composition is strongly influenced by rock type. On highly acidic scree, by far the most important colonist is the Parsley Fern. It is a pioneer of unstable scree, and unlike other scree ferns, is capable of extending beyond the recesses to form clumps, which then bring a degree of stability by growing over the tops of scree boulders. This may then allow other species such as Male-fern (*Drypteris filix-mas*), Scaly Male-fern (*D. affinis*) and Mountain Male-fern (*D. oreades*) to flourish. The leaching that takes place under high rainfall, means that even some of the screes under base-rich cliffs now support acid-loving plants, and this is clearly the case in parts of Cwm Idwal. Truly base-rich screes on the other hand, in places such as Craig Las on the northern slopes of Cadair Idris, are the nearest things to a fern paradise in Snowdonia. Species of *Asplenium* are often particularly abundant, but Lemon-scented Fern (*Oreopteris limbosperma*), Beech Fern, Oak Fern, Hard Shield-fern, Brittle Bladder-fern, and several buckler ferns may all occur.

WOODLAND FERNS

Woodland is a habitat for more fern species than any other. This is partly because ferns have the ability to grow in heavy shade, but the high humidity, the shelter, and the absence of competition from flowering plants, are all important factors. The rocky oakwoods of Snowdonia have varied ground floras, but typically include stands of tall species such as Male-fern, Lady-fern and Broad Buckler-fern, especially where the grazing pressure is low. Heavily grazed woods such as Coed Cymerau tend to have a grass-dominated field layer. Bracken (*Pteridium aquilinum*) is also likely to be a feature of many of these woodlands, even though we now tend to think of it more as a plant of open hillsides. The magnificent Royal Fern would once have been present here, perhaps in abundance, but it has become rare as a result of its being so eagerly sought by collectors, especially in Victorian times. Pre-1800 travellers, for example, mention seeing it in the vicinity of the wooded gorge at Aberglaslyn – a place where it is certainly not found today. None the less, there are still some fine specimens in the wooded gorge of Ceunant Llennyrch.

The soil in Snowdonia's woodlands is usually acidic, but where conditions become slightly more basic, especially close to woodland streams, the fern flora can be particularly luxuriant. In boulder-strewn dingles in Coed y Rhygen, for example, it is possible to see many different ferns growing together including Lemon-scented Fern, Beech Fern, Oak Fern, and Common Scaly Male-fern (*Dryopteris affinis* subsp. *borreri*). Many of the trees in these woodlands are likely to be festooned with polypody ferns (*Polypodium* spp.) – plants with the distinction of being Britain's only vascular epiphytes (plants that grow on other plants but take nothing from them). There are three British species of which two, the Common Polypody (*P. vulgare*) and the Intermediate Polypody (*P. interjectum*), are common. The Hard Fern is more characteristic of moist banks – on the steep wooded slopes of Ceunant Glanrafon near Rhyd Ddu, for example, or at sites like Coed Victoria near Llanberis and Dinas Emrys in Nantgwynant, where there are rocky outcrops. Wilson's Filmy-fern is also abundant in such places – unlike its close relative the Tunbridge Filmy-fern, which is far rarer, even though it has similar habitat requirements. Both are classed as oceanic species, indicating a distribution limited to the Atlantic zone and not extending eastwards into central Europe. They also tend to be restricted to the western side

Riverside ferns on the Afon Rhaeadr Fawr.

of Britain, and have a close association with the humid woodlands of the Atlantic coast. Coed Ganllwyd, for example, contains the Tunbridge Filmy-fern, and Coed Gors y Gedol is one of the few localities for another oceanic species, the Hay-scented Buckler-fern. The Narrow Buckler-fern (*Dryopteris carthusiana*) is not especially oceanic, occurring as far east as Siberia, but it is scarce in Snowdonia occurring in only a few marshy woodlands such as Coed Llechwedd near Harlech. Other woodlands, such as Coed Aber Artro in Meirionnydd and the Betws-y-Coed/Gwydir area in Dyffryn Conwy, contain lime-loving species such as the Hard Shield-fern, the Soft Shield-fern (*Polystichum setiferum*) and the Hart's Tongue.

Other special places for ferns are oak-clad ravines such as Ceunant Cynfal, Ceunant Dulyn, Ceunant Llennyrch and the aforementioned Coed Aber Artro. The humid, shady atmosphere of some of these places reminds one of a tropical rain forest, and the ferns can reach lavish proportions. Great stands of Broad Buckler-fern and other common species cling to the valley sides, and where the soil is more acidic, the Hard Fern is likely to be dominant. The rich fern flora of Ceunant Llennyrch also includes Tunbridge Filmy-fern and Hay-scented Buckler-fern.

WETLAND FERNS

Very few ferns are fully aquatic but there is one species that occurs in Snowdonia. It is the very un-fern like Pillwort (*Pilularia globulifera*), which can easily be overlooked or mistaken for young rush plants. It tends to colonise the muddy shorelines of pools and lakes where competition from the other plants is low. It has decreased markedly in recent times however, and can now only be found at a few sites such as Llyn Idwal and Bwlch Sychnant (Sychnant Pass). The Royal Fern is at home in wetland habitats, and occasionally occurs on marshy ground adjacent to damp woods and lakes. Llyn Ty'n y Mynydd has perhaps the strongest colony, with more than 100 plants. The Marsh Fern no longer exists in Snowdonia, in spite of the apparent availability of suitable habitats.

Royal Fern (*Osmunda regalis*) near Llyn Ty'n y Mynydd.

FERNS OF HEATH AND MOORLAND

Heathland and moorland are typically dominated by low-growing shrubby plants of the heather family. The difference between them is that heathlands occur on shallow peat or mineral soil at comparatively low altitude in valleys, cwms and foothills, whereas moorlands occupy higher ground, often with deep accumulations of peat.

Bracken is by far the most abundant and widespread fern on well-drained heathland, but Hard Fern can also be quite common, especially on more humid sites. Lemon-scented Fern occurs in sheltered pockets at moderate or high altitude – up to 900 m in some cases – and can be locally dominant along the fringes of steep mountain streams, even reaching into the valley bottom – as in Nant Ffrancon. Heathlands are also a habitat for various buckler ferns. Common Scaly Male-fern can be locally common, especially on some of the steeper well-drained sites, usually as the subspecies *borreri*. Narrow Buckler-fern grows on damper heathland sites, but is often not recognised because of its similarity to the much commoner Broad Buckler-fern. The two species grow together in the areas around Bwlch Sychnant and Cwm Dulyn (Llanllyfni).

GRASSLAND FERNS

Bracken is overwhelmingly *the* fern of upland acidic grassland, but other species can occur in some of the more base-rich or neutral pastures. In Nant Bwlch-yr-Haiarn in upper Dyffryn Conwy for example, it is possible to find the Adder's-tongue (*Ophioglossum vulgatum*), and the Moonwort. The latter species has recently been recorded in Cwm Glas Mawr and Cwm Du'r Arddu. Both appear to have been widely known as grassland plants during the Medieval Period, and mystical properties were ascribed to them – probably because of their unusual forms and the fact that they do not show themselves consistently from year to year. Adder's-tongue was thought to poison the grass among which it grew, and to cause injury to any cattle that ate it. Its scientific name, *Ophioglossum*, is derived from two Greek words meaning serpent's tongue. The key-like shape of the moonwort was thought capable of opening locks and undoing shackles, and alchemists believed it could turn mercury (quick-silver) into real silver. The name 'Moonwort' probably comes from the resemblance of the cup-shaped pinnae to the crescent moon.

THE FERNS OF MAN-MADE HABITATS

Despite their seeming wildness, the uplands of Snowdonia, even up to the highest summits, bear the marks of man's intrusion. Sturdy dry-stone walls straddle the remotest ridges, piles of slate quarry debris scar many of the lower mountain slopes and waste from abandoned copper and lead mines is widespread. Surprisingly perhaps, these artificial habitats are now known to be important refugia for ferns.

Four species of fern, the Black Spleenwort (*Asplenium adiantum-nigrum*), Common Maidenhair Spleenwort (*Asplenium trichomanes* subsp. *quadrivalens*), Rustyback Fern (*Asplenium ceterach*), and the appropriately named Wall-rue (*Asplenium ruta-muraria*), grow more abundantly and luxuriantly on walls than in their natural rocky habitats. A hybrid between Black Spleenwort and Forked Spleenwort was collected in Llanberis Pass by the Rev. Thomas Butler in 1870, and is still known as the Caernarvonshire Fern (*Asplenium* x *contrei*), even though there are no recent records for Snowdonia. The closely related Lanceolate Spleenwort (*Asplenium obovatum* subsp. *lanceolatum*) is mostly a coastal species, but it has been recorded in rock crevices as far inland as Capel Curig in the heart of Snowdonia. It will also grow on walls, and has been recorded on a garden wall at Nantmor near the Aberglaslyn Pass. In a more natural setting, it grows on the cliffs between Tan y Bwlch and Aberglaslyn overlooking the vast expanse of reclaimed land stretching seaward to Porthmadog. Prior to the construction of the embankment (known locally as The Cob) between Porthmadog and Minffordd by William Madocks in the early 1800s, tidal waters would have extended right up to the cliffs of Aberglaslyn. Further south, in the seaside town of Abermaw (Barmouth), the fern is so common on the rocks and walls of the town that it used to be called the Barmouth Fern. Slate tips are generally hostile to plants including ferns because of their instability, but there is one species, the Parsley Fern, that is so well adapted to them that it achieves higher densities than on many natural screes. One of the most improbable artificial habitats for ferns is the highly toxic waste that was left in the wake of the lead mining industry. Spoil heaps of this material can be found scattered throughout Coed Gwydir, and despite levels of lead and zinc that would be lethal to most plants, a few species flourish on them. The Forked Spleenwort is actually more abundant on these sites than in any natural habitat. Black Spleenwort and Lady-fern also occur in lesser quantities.

BRACKEN

When woodland was the dominant vegetation in Snowdonia and in other parts of Britain, Bracken would have been mainly confined to shaded clearings and woodland edges, and spore counts taken from lake-bed cores confirm that up until about 7000 years ago, it was a relatively uncommon plant. It began to increase, though, as forest clearance by humans took effect, and reached a peak about 5000 years ago. Further expansion came only in historic times following a renewed assault on the forest, and the trend has continued to the present day, perhaps even accelerating in the last 50 years as a result of changes in farming practice. Part of the reason is that Bracken is no longer cut as a crop (for animal bedding or for use on haystacks), but changes in livestock have also made a difference. Where at one time cattle grazed the hills and controlled the Bracken by trampling, the sheep that replaced them make less impact. The fact that sheep avoid Bracken while eating almost everything else means that Bracken gains a competitive advantage, allowing it to spread at the expense of species that are grazed – grasses and heather especially.

Bracken in autumn on the hills above Nant Ffrancon.

Bracken has now become so abundant that it forms an important element in the landscape, and with its seasonal colour changes – deep green in summer, russet-gold in autumn and rust-brown in winter and spring – it undoubtedly adds visual variety. It is a curious fact though, that because its method of spread is entirely by rhizomes, the large tracts that we now see in places such as Nant Ffrancon and on Cadair Idris may well consist of a single plant, possibly centuries old.

Bracken's success has been attributed to its high resistance to disease, its low palatability, its toxic effects on competing plant species, its pronounced vegetative longevity, its ability to tolerate fire, and its wide climatic and edaphic tolerance. With such attributes, it is hardly surprising that it is so widespread, and that its spread continues – currently at a rate between 1% and 3% per annum. If unchecked, this would result in a doubling or even a trebling of its extent over the next millennium. Such expansion would eventually be limited by the lack of suitable soils – deep well-drained mineral soils are essential – and by its inability to live above about 600 m. Even so, the losses to agriculture could be serious, as the following farmer's lore indicates.

> *Aur dan y rhedyn*   *Gold under the Bracken*
> *Arian dan yr eithin*   *Silver under the gorse*
> *Newyn dan y grug*   *Famine under the heather*

In fact, Bracken spread is not just a Welsh or British problem, but a worldwide one, earning the plant a reputation as the world's worst weed.

CONSERVATION

The fern-collecting mania lasted throughout the Victorian age, reaching its peak during the 1860s, but the gathering of plants in general continued throughout the 19th century, well into the 20th, and in some instances still goes on. Happily though, attitudes generally are now more enlightened, and the camera has to a great extent replaced the collector's vasculum as part of a botanist's equipment. The introduction of the Wildlife and Countryside Act and other associated legislation has helped safeguard the future of some of our rarest ferns, with species such as Killarney Fern, Alpine Woodsia and Oblong Woodsia now protected by law. The Killarney Fern and Oblong Woodsia are also Red Data Book species listed by the World Conservation Union as being vulnerable on a global scale. Several ferns found in Snowdonia are classed as 'Nationally Scarce', which means they occur in fewer than 100, 10 x 10 km squares nationwide. Forked Speenwort, Lanceolate Spleenwort, Marsh Fern and Pillwort are examples. They are protected mainly by virtue of their presence on statutary sites, particularly Sites of Special Scientific Interest and National Nature Reserves.

FURTHER READING

Allen, D. E. 1969. *The Victorian Fern Craze: A History of Pteridomania*. London.

Allen, D. E. 1993. *The Victorian Fern Craze: Pteridomania Revisited*. Fern Horticulture: past, present and future perspectives.

Condry, W. 1966. *The Snowdonia National Park*. The New Naturalist. Collins, London.

Griffith, J. E. 1894/5. *The Flora of Anglesey and Carnarvonshire*, Bangor.

Hutchinson, G. & Thomas, B. A. 1996. *Welsh Ferns Clubmosses, Quillworts and Horsetails*. Cardiff.

Jones, D. 1996. *The Botanists and Guides of Snowdonia*. Llanrwst.

Newman, E. 1854. *A History of British Ferns*. London.

Page, C. N. 1988. *Ferns – their habitat in the British and Irish Landscape*. The New Naturalist. Collins, London.

Raven, J. & Walters, M. 1965. *Mountain Flowers*. London.

Ray, J. 1724. *Synopsis Methodica Stirpium Britannicarum*. The Ray Society, London 1973.

Roberts, R. H. 1979. The Killarney Fern, *Trichomanes Speciosum*, in Wales. *Fern Gazette*, 12 (1): 1-4.

Secord, A. 1994. Science in the Pub: Artisan Botanists in Early Nineteenth-Century Lancashire. *British Journal for the History of Science*, xxxii. 269-315.

Woodhead, N. 1933. *The Alpine Plants of the Snowdon Range*. Alpine Garden Society Publication.

CLUBMOSSES AND QUILLWORTS

PETER RHIND

INTRODUCTION

The clubmosses and quillworts are the living representatives of a group of ancient vascular plants (the lycopods) with an evolutionary line extending back over 400 million years to the Silurian Period. There are eight species of clubmoss in Britain and five of them, Alpine Clubmoss (*Diphasiastrum alpinum*), Fir Clubmoss (*Huperzia selago*), Lesser Clubmoss (*Selaginella selaginoides*), Marsh Clubmoss (*Lycopodiella inundata*) and Stag's-horn Clubmoss (*Lycopodium clavatum*) occur in Snowdonia, although the Marsh Clubmoss is very rare. A sixth species, the Interrupted Clubmoss (*Lycopodium annotinum*), has also been recorded, but not for over a century. It was last seen above Llyn y Cŵn in the 1690s and on the Glyderau up until the start of the 19th century. Quillworts are represented by three species in Britain, but only two, Quillwort (*Isoetes lacustris*) and Spring Quillwort (*Isoetes echinospora*), occur in Snowdonia, and the latter is scarce.

Clubmosses and quillworts have a relatively simple structure, and it is not surprising that early botanists classified clubmosses with the mosses. Unlike mosses, though, both groups have a distinct vascular system for transporting water and nutrients, albeit simpler in structure than in more modern plant groups such as flowering plants. It mainly takes the form of a central cylinder (known as protostele) and appears to be similar to the vascular systems found in some of the earliest plants to colonise land. The fossil record shows that such structures developed during the Silurian Period and some of the plants of this period, such as *Zosterophyllum*, seem to have had much in common with the clubmosses of today. A fossil plant called *Baragwanathia* from the upper part of the Silurian may indeed have been the earliest lycopod. By the Devonian Period, some 50 million years later, taxa such as *Asteroxylon* and *Drepanophycus* had evolved that were almost indistinguishable from some modern day clubmosses such as the Fir Clubmoss. However, unlike living clubmosses, these very early species did not have true roots, and their leaves appear to have been little more than simple, unvascularised outgrowths which probably developed to increase the photosynthetic surface of the plant, but were not leaves in the modern sense.

The lycopods split into two major groups during their early development; one remained herbaceous and includes all of the living species, while the other (the lepidodendrid) became woody and treelike and gave rise to one of the dominant plant groups of the coal-forming forests that developed some 300 million years ago during the Carboniferous Period. Some of the herbaceous species of this period, such as *Selaginellites*, were very similar to the Lesser Clubmoss.

Like the ferns, the clubmosses and quillworts have life cycles with two distinct generations: a large highly structured one that produces spores (the sporophyte), and a tiny one that produces the male and female gametes (the gametophyte). The sporophyte is the plant that we normally see in the field, and produces spores on structures known as sporangia located on or near the base of specialised leaves known as sporophylls. In clubmosses these are grouped either in distinct terminal cones known as strobili (e.g. Stag's-horn Clubmoss) or in fertile zones distributed along the stem (e.g. Fir Clubmoss), whereas quillworts produce a flush of sporophylls followed by a flush of sterile leaves. In what is assumed to be the more primitive of the two clubmoss families, the Lycopodiaceae, all the species have sporangia of similar size (homosporous). They differ from the species in the lesser clubmoss family (the Selaginellaceae) and the quillwort family (the Isoetaceae), whose sporangia are of two distinct sizes: megasporangia (large) and microsporangia (small). Such species are described as heterosporous.

Homosporous species produce bisexual, free-living gametophytes, but unlike the gametophytes of ferns, they are commonly subterranean, relying for their nutrition on an association with a mycorrhizal fungus. They ultimately develop male and female organs (the antheridia and archegonia respectively), but sometimes take as much as 15 years to complete the process. The antheridia produce motile sperm, which finally fulfil their function by swimming to the archegonium through a film of water.

Heterosporous species, on the other hand, have a slightly different life cycle, in that their gametophytes are unisexual. At maturity the male gametophytes simply consist of a minute antheridium, which develops within the spore of the microsporangium. This eventually ruptures, allowing the sperm to swim free. The female gametophyte is also greatly reduced and develops within the spore of the megasporangium. Eventually this also ruptures, exposing the archegonium. The gametophyte generation in these species is therefore not only reduced to its bare essentials, it also gains much support from the sporophyte generation. The seed plants represent the extreme expression of this development, in that their gametophyte phase is greatly reduced, and the young sporophyte or embryo develops within the seed of the previous sporophyte generation before dispersal. The formation of two types of spore, as seen in some of the lycopods, is therefore thought to represent a stage in the evolutionary process that led to the development of the seed plants.

Clubmosses are also capable of spreading by means of creeping stems or stolons but, among the British species, only the Fir Clubmoss produces bulbils (small bulblike structures). These flattened structures are produced near the base of sterile leaves, and sometimes form a crown at the top of the stem. Bulbils are often more successful as a means of vegetative spread than stolons, particularly in patchy environments. However, all the species rely on spore production for wide dispersal, and as a means of colonising newly created bare-ground habitats.

THE HISTORY OF CLUBMOSS AND QUILLWORT RECORDING IN SNOWDONIA

Although the earliest written records for Snowdonian clubmosses come from various sources, all of them date from the 17th century. For Alpine Clubmoss, there is a record in *Phytologia Britannica* (Howe,1650), with the location given as 'on the top of Snowdon'. Marsh Clubmoss is recorded in *Camden's Britannia* (Gibson, 1695) at 'Capel Ceirig' [sic], while Fir Clubmoss and Stag's-horn Clubmoss are listed as occurring 'On Snowdon and other high mountains of Wales'. Lesser Clubmoss is recorded from Snowdon in the *Catalogus Plantarum Angliae* (Ray, 1670), and Interrupted Clubmoss is described in the *Synopsis Methodica Stirpium Britannicarum* (Ray, 1690) as occurring 'Above Llin y Cŵn' [Llyn y Cŵn] (a small lake just below the summit of Glyder Fawr). This very rare species had been recorded there by Edward Lhuyd (of Snowdon Lily fame) and on the Glyderau in 1726 by the German botanist Johann Jacob Dillenius and by Samuel Brewer. This record was repeated after a gap of over 100 years (1832) when the eminent botanist Charles Cardale Babington and his friends refound it at this same site. This seems to have been shortly before its demise, however, for the only later record was that of William Wilson in 1836. J. E. Griffith in his *Flora of Anglesey and Carnarvonshire* (1894/5) states, 'I am afraid this is extinct in Carnarvonshire. It used to grow on the Glyders, but I have failed to find it for years now'.

John Ray provides the first record of Quillwort in Snowdonia (also the first for Wales) in his *Synopsis Methodica Stirpium Britannicarum* (1690). The plant had been found by Edward Lhuyd in Llyn Bach on Snowdon. Some 37 years later in 1727, Samuel Brewer, during his botanical exploration of Snowdonia (see *Samuel Brewer's Diary* – Report of the Botanical Society and Exchange Club for 1930 by H. A. Hyde) also found it in Llyn Dwythwch and Llyn Ogwen, and today it is known to occur in at least twelve of Snowdonia's lakes. Spring Quillwort was not discovered until 1862 when Charles Cardale Babington, Professor of Botany at Cambridge University, found it in Llyn Padarn. The record, also a first for Wales, was published in the *Journal of Botany* (1863) and in his diary (see *Memorials Journal and Botanical Correspondence of Charles Cardale Babington*, 1897) he describes finding '*Isoetes echinospora* in the lower lake [Llyn Padarn] near Ynys on the left bank'. During the following days he also found it in Llyn y Cŵn below Glyder Fawr and in the Afon Rhyddallt at Cwm y Glo, near to where the river issues from Llyn Padarn, but today it is known to be far less common than Quillwort.

Studies of spores from sediment cores from various parts of Snowdonia show that quillworts, like other lycopods, have been present in Snowdonia for many thousands of years. In the period beginning with the retreat of glaciers about 10,000 years ago, clubmosses were among the pioneer colonisers of the newly exposed soils. All our present-day species appear to have been much more widespread at that time, in both lowland sites such as Nant Ffrancon, and upland ones such as the area around Llyn Dwythwch, Llyn Llydaw and Capel Curig. At

Llewesig near Capel Curig, for example, Fir Clubmoss, Interrupted Clubmoss, Lesser Clubmoss and Stag's-horn Clubmoss were all present during this period, whereas the nearby site of Gors Geuallt contained all the above with the exception of Stag's-horn Clubmoss. The Interrupted Clubmoss, now absent from Snowdonia, seems to have been rather common at that time, to the extent that in some areas – around Nant Ffrancon and Llyn Dwythwch for example – it may have been the only species present.

During this late glacial period, it seems that Snowdonia's newly formed lakes, such as Llyn Dwythwch and the now lost lake of 'Llyn Nant Ffrancon', were rapidly colonisd by quillworts. Spring Quillwort was initially the only species, but Quillwort seems to have appeared perhaps 1000 or 2000 years later as climatic warming continued. Spring Quillwort has now disappeared from Llyn Dwythwch, but the spore record indicates that it was the more common of the two species well into the early postglacial period. That situation is now reversed, suggesting perhaps that Spring Quillwort exists in Snowdonia only as a relict of a former cold period.

During the early postglacial period much of Snowdonia became covered in tundra vegetation, and while some places such as Nant Ffrancon and Llyn Dwythwch were dominated by Dwarf Birch (*Betula nana*), others, like the area around Llyn Cororion, had a kind of dwarf-heath vegetation dominated by Crowberry (*Empetrum nigrum*). Both types included a large herbaceous element and were a habitat for Fir Clubmoss, Interrupted Clubmoss and Lesser Clubmoss. However, by about 9500 years ago, woodland had begun to spread over much of the lowland and semi-upland areas. It was initially dominated by birch but, by about 8000 years ago, oak had become well established as the dominant species, and pine forest had colonised much of the uplands. This period must have brought about a great contraction in clubmoss habitat, and it is likely that most species became restricted to the high alpine slopes and mountain tops. Only Interrupted Clubmoss may have continued to grow in the pine forests, as it does today in the native pine forests of Scotland. The fact that it also grows in the mixed pine and birch forests of continental Europe perhaps means that it would also have found a niche in the early birch forests of Snowdonia.

Forest clearance by Neolithic farmers provided conditions that allowed some of the clubmosses to expand their range to lower altitudes. In the early stages, perhaps 5000 years ago, the practice was for small patches of woodland to be cleared and used for crops for only a few years before they were abandoned and the trees allowed to return. The impact on clubmosses at that time would have been slight, but as clearance became more extensive, and grazing by domestic stock began to hinder the natural re-generation of woodland, some clubmosses would have benefited. The semi-natural mountain grasslands in particular, would have become more widespread, providing a habitat for clubmosses such as the Alpine Clubmoss, Fir Clubmoss and Stag's-horn Clubmoss. Only the Interrupted Clubmoss seems to have suffered a decline, but that may have been for climatic reasons.

IMPORTANT HABITATS FOR CLUBMOSSES AND QUILLWORTS IN SNOWDONIA

The three most common clubmosses, Alpine Clubmoss, Fir Clubmoss and Stag's-horn Clubmoss, are mainly associated with acidic mountain grassland, heathland and moorland, although the last species requires traces of basic salts. Nevertheless, all three species have been found growing together in Clogwyn Du'r Arddu. They are all locally common at high altitude, but rarely occur below an altitude of 300 m. On some alpine slopes – the north-facing slopes of Elidir Fach, for instance – Alpine Clubmoss can occur in great abundance to form, in some cases, an Alpine Clubmoss turf. In other areas it may be associated with the Fir Clubmoss, or form mosaics with Crowberry. Both Alpine Clubmoss and Fir

Alpine
Clubmoss
(*Diphasiastrum
alpinum*) on
Elidir Fach.

Fir Clubmoss
(*Huperzia selago*)
on Moel Hebog.

Clubmosses are also common in mountain-top communities, where they may be associated with alpine flowering plants such as Stiff Sedge and Dwarf Willow. Stag's-horn Clubmoss, on the other hand, is rarely seen above 800 m, and perhaps because of sensitivity to grazing, is not usually found on open grassland. It is more likely to be encountered among rocks in boulder fields or on block screes in places like Cwm Idwal or below Gallt yr Ogof in Nant y Benglog. Rock crevices and ledges are also important habitats for clubmosses. Both the Fir Clubmoss and the Lesser Clubmoss can be found in such places, the latter less commonly than the former because of its requirement for base-richness and moisture. The crags of Clogwyn Du'r Arddu and Cwm Glas are two of its localities. These are also the conditions required by many of the arctic-alpine flowering plants, so Lesser Clubmoss is often associated with rarities such as Northern Rockcress and Alpine Meadow-rue.

Stag's-horn Clubmoss (*Lycopodium clavatum*) on Elidir Fach.

In contrast, the rarest of all our clubmosses, the Marsh Clubmoss, is a species of peat bogs and wet acid heaths. It never occurs in quantity, but a few scattered plants can be found, for example, in Nant y Benglog, and it has recently been discovered in Cwmffynnon. Finally, prior to its disappearance from Snowdonia, the Interrupted Clubmoss would have been found growing in mountain grasslands and moorlands. In Scotland it also occurs in native pine forest.

In contrast to the clubmosses, the quillworts are aquatic and restricted mainly to alpine and sub-alpine lakes. Quillwort, for example, has been recorded in many of Snowdonia's lakes, including Llyn Ogwen, Llyn Idwal, Llyn y Cŵn, Llyn Padarn, Llyn Peris, Llyn Llagi and Llyn yr Adar, but Spring Quillwort is

Lesser Clubmoss (*Selaginella selaginoides*) on Clogwyn Du'r Arddu.

Marsh Clubmoss (*Lycopodiella inundata*) in Nant y Benglog.

much less common. It has been recorded, though, in the Llanberis lakes, Llyn Idwal, Llyn Crafnant and Cwm y Glo. Most of the lakes with quillwort populations are low in nutrients, and it was once thought that they required such conditions. The finding of Spring Quillwort in two nutrient-rich lakes on Ynys Môn (Anglesey) is evidence, though, for an alternative suggestion that Quillworts may be excluded only by competition from plants that respond better to high nutrient levels.

CONSERVATION

Two species, Spring Quillwort and Marsh Clubmoss are classed as Nationally Scarce in Britain, which means that over the country as a whole they occur in fewer than 100 10 x 10 km squares. As the legislation stands, this would not normally bestow any additional protection, but because they are listed as features of interest within the Snowdonia Site of Special Scientific Interest, they do benefit from protection under the Wildlife and Countryside Act. In addition, all clubmoss species are protected under Annex Vb of the EC Habitat and Species Directive (92/43/EEC), which states that these are 'Plant species of Community interest whose taking in the wild and exploitation may be subject to management measures'. It mainly relates to the taking of large numbers of specimens for commercial gain, but where such management measures are necessary they may include restricting access to certain areas or prohibiting the taking of specimens.

ADDITIONAL READING

Jermy, A. C. 1993. British Clubmosses. *British Wildlife*, 4:216-220.

Page, C. N. 1988. *Ferns – their habitat in the British and Irish Landscape*. The New Naturalist. Collins, London.

Seddon, B. 1965. The occurrence of *Isoetes echinospora* in eutrophic lakes in Wales. *Ecology*, 46: 747-748.

HORSETAILS

PETER RHIND

INTRODUCTION

Like clubmosses, horsetails are a very ancient group with a fossil record traceable back some 300 million years to the Devonian period. In terms of species diversity and abundance, they reached their peak in the late Palaeozoic era, roughly 300 million years ago, when some species grew to heights of 15 m or more. During the late Devonian and Carboniferous periods, they combined with giant clubmosses to form dense swamp forest, the remains of which eventually became the coal that was to play such a vital role in the industrial revolution. Today only one herbaceous genus, *Equisetum*, survives, with just 23 species worldwide. The majority of these occur in the Northern Hemisphere, and five of them, Field Horsetail (*Equisetum arvense*), Water Horsetail (*E. fluviatile*), Marsh Horsetail (*E. palustre*), Wood Horsetail (*E. sylvaticum*) and Great Horsetail (*E. telmateia*) can be seen in Snowdonia. It is also possible to see the hybrid between *E. arvense* and *E. fluviatile*, known as the Shore Horsetail (*E. x litorale*). This, as one would expect, has features in common with both parent plants.

Although primitive, horsetails have a fairly elaborate vascular system with small vascular bundles distributed through the stem cortex rather than concentrated in a central core as in the clubmosses. In this respect, horsetails are much more akin to higher plants. They are unique, however, in their conspicuously jointed stems and their small scale-like leaves arranged in whorls. If branches are present these also occur in whorls and alternate with the leaves. Some species, therefore, look like miniature Christmas trees. Surprisingly, though, such whorled structures are rather rare in nature, although they have developed independently in the flowering plant family Rubiaceae (the bedstraws), and in the algal family Charophyceae (the stoneworts). The aerial stems arise from branching underground rhizomes and, although they may die back during the winter, the rhizomes are perennial.

An unusual feature of horsetails is their ability to incorporate silica into their structures. Their surfaces are covered with minute silica spines that make them rough to the touch, and give them utility as a form of very fine sandpaper, once employed as a polishing medium for wood and pewter. Although all the species possess this property, some species are more abrasive than others, and Dutch

Rush (*Equisetum hyemale*) seems to be the most highly regarded of all, although it is not a species that we have in Snowdonia.

Like other lower plants, horsetails reproduce by spores and have alternating sexual and asexual generations. Spores are produced by the so-called sporophyte generation, which is the plant we normally see in the field. Each plant produces umbrella-like structures known as sporangiophores (sporangia-bearing branches), which develop in clusters to form cones (strobili). These are produced at the top of the stem, but some species, such as the Field Horsetail and the Great Horsetail, produce special fertile stems, which are devoid of chlorophyll, and usually develop early in the season. They are known as deciduous-stemmed horstails. Other species, such as the Marsh Horsetail, simply produce cones at the top of normal green stems. It is interesting to note that the Great Horsetail is not only our largest native horsetail, but also the largest deciduous-stemmed horsetail in the world.

In contrast to some clubmosses, horsetails have sporangia of uniform size (homosporous) which, when mature, split to release numerous spores. These germinate to produce the gametophyte or sexual generation: a tiny free-living plant no bigger than a pinhead. These eventually develop sex organs known as antheridia (male) and archegonia (female), but water is required for the multi-tailed sperm to swim to the female organ and complete the life cycle.

RECORDING HISTORY

Although it is not possibly to say which species, horsetails have been present in Snowdonia since at least the late glacial period some 12,500 years ago. Spores from this period have been found in glacial deposits at Llyn Llydaw, Glanllynnau (near Criccieth) and Cors Geuallt (near Capel Curig), and spores dated from the immediate postglacial period have been found in Nant Ffrancon. The horsetails at the Glanllynnau site are likely to have been Water Horsetail since they were found associated with sedges growing in what appear to have been swampy conditions, but as the climate during this period was extremely cold, there is a strong possibility that arctic-alpine species, such as the Shady Horsetail (*E. pratense*), were also growing in the area at the time.

Curiously, very few of the botanical pioneers of Snowdonia refer to horsetails in their published records, but this may relate to their near absence in the uplands. One of the earliest records appears in *Samuel Brewer's Diary* (see Report of the Botanical Society and Exchange Club for 1930 by H. A. Hyde), where in 1727 he describes seeing an *Equisetum* (possibly *E. telmateia*) in a meadow in Nant Ffrancon, which he thought differed from all he had seen previously. After that there appear to be few references to horsetails, and most of them refer to the rarer Great or Wood Horsetail. For example, J. E. Griffith lists Wood Horsetail for Felin Hen and Nant Ffrancon in his *Flora of Anglesey and Carnarvonshire* (1894), and P. M. Benoit and M. Richards in their *Contribution to a Flora of Merioneth* (1961) list it for Blaen-y-Cwm, Llangower, Esgair-Gawr (near Dolgellau), near Hafod-y-Meirch, the banks of Wnion near Dolserau,

Penrhyn Gwyn, and Tyn-y-Coed near Dinas Mawddwy. In the latter flora, Great Horsetail is listed for Talybont Wood and Llety Lloegr, a ditch below Llwynwcws (near Llanaber), and on the roadside at Rhoslefain. Other species were probably deemed to be too frequent for authers to list specific sites and are usually just described as common throughout the area.

IMPORTANT HABITATS FOR HORSETAILS IN SNOWDONIA

In most cases the common names of horsetails give an indication of their habitat requirements, including grasslands (Field Horsetail), marshlands (Marsh Horsetail), woodlands (Wood Horsetail), and open water (Water Horsetail), although these species are not necessarily restricted to these habitats. The Field Horsetail is the least moisture demanding of all the species, and the only one that can become established as a weed of cultivated areas and waste ground. It can even penetrate tarmac and often occurs on road edges in Snowdonia.

UPLAND HABITATS

Of the British horsetails only two, the Shady Horsetail (*E. pratense*) and Variegated Horsetail (*E. variegatum*) are classed as true mountain species. Both can be found in the mountains of Scotland and northern England, but they are regrettably absent from the mountains of Snowdonia. The latter species however, is commonly found growing in dune slacks in Wales, and in this respect it is comparable with certain flowering plants such as Thrift (*Armeria maritima*) that occur in both coastal and mountain habitats, but are absent from intervening habitats. But why this species is absent from the mountains of Snowdonia is a mystery. Unbranched forms of the Field Horsetail have occasionally been recorded in wet upland habitats, where they have been mistaken for the Variegated Horsetail.

HEATHLAND AND MOORLAND

Heathland and moorland up to an altitude of about 450 m, constitute two of the more important habitats for horsetails, especially where damp mineral soil banks are exposed, or where there is base-rich seepage. Horsetails found in these habitats may include Field Horsetail, Water Horsetail, Marsh Horsetail and Wood Horsetail, though the last species is thought to be a survivor from former forest cover, and able to tolerate these more open conditions only in the north and west of Britain where frequent cloud cover and regular precipitation mimic the more humid and shady conditions found in woodlands. It is generally uncommon, but has been recorded in heathland adjacent to Llyn Cowlyd. The other three species are all relatively common, and sometimes form pure stands given the right conditions.

Wood Horsetail (*Equisetum sylvaticum*) on the Carneddau near Llyn Cowlyd.

WETLAND AND AQUATIC HABITATS

In marshy conditions, especially where there is some base enrichment, the Marsh Horsetail tends to be the main species. It is common throughout Snowdonia, sometimes accompanied by scattered shoots of the Water Horsetail, but as our most aquatic horsetail, the Water Horsetail is more at home growing in nutrient-poor ponds and lakes. Extensive stands occur, for example, in Llyn Bochlwyd and Llyn Idwal, and it has been recorded in the flooded quarry below Marchlyn Bach Reservoir. It can also tolerate the shady conditions found in alder and willow carr, and it has become the dominant species in certain marshy areas such as Cors Gyfelog near Pant Glas. The hybrid between Water Horsetail and Field Horsetail, known as Shore Horsetail is normally found in places that are too wet for Field Horsetail and too dry for Water Horsetail. It typically occurs in ditches, fens, damp grasslands and

Marsh Horsetail (*Equisetum palustre*) on the Carneddau.

by lake and river margins, and is thought to be more widespread than the few records for Snowdonia suggest, due to confusion with its parent plants. However, it has been recorded on a number of occasions in the drainage ditches adjacent to the Afon Glaslyn near Minffordd. Both the Great and Wood Horsetails may also occasionally be found on open marshy ground. Wood Horsetail, for example, has been found growing in mire communities alongside species such as Dioecious Sedge (*Carex dioica*) in Cwm Idwal and Bottle Sedge (*Carex rostrata*) and Purple Moor-grass (*Molinia caerulea*) at Tir Stent near Dolgellau.

Reproductive cone of Great Horsetail (*Equisetum telmateia*) in Coedydd Aber.

WOODLANDS

Most British horsetails can occasionally be found in woodland, but none of them is restricted to it. The Wood Horsetail, as its name suggests, is the species most closely associated with woodlands, but it is comparatively rare in Snowdonia. Only in the more northern areas of Britain does it become a frequent component of woodlands, and most of the records for Snowdonia are from sites outside woodlands, but it has been found in the woods near Rowen in Dyffryn Conwy. The Great Horsetail is also often associated with woodlands, but, in contrast to Wood Horsetail, has more of a southern distribution in Britain and is absent from much of Scotland. The species has a liking for water-logged, clay soils, especially where there is a degree of base-richness, and there is a particularly large stand in Coedydd Aber. Other sites include the valley woods in Cwmgwared on Bwlch Mawr.

GRASSLANDS

The Field Horsetail and Marsh Horsetail occur in a variety of grasslands in Snowdonia, with Field Horsetail preferring the drier types, while Marsh Horsetail is more at home in rush pasture. The Marsh Horsetail can also be found in some of the upland grasslands growing alongside species such as Mat-grass, Sheep's Fescue and Heath Rush, and has been recorded, for example, in

Cwm Eigiau and in fields close to Llyn Coedty on the Carneddau. On rare occasions, Great Horsetail and Wood Horsetail may also be found in grasslands. In fact, the former species was first recorded in Wales by Samuel Brewer in 1728 'from ye meadows at Nant Ffrancon'. However, they tend to be restricted to damp shady banks and streamside fields. In his book, *The Snowdonia National Park* (1966), William Condry mentions having seen Wood Horsetail in a streamside field in the Hirnant Valley near Rhos-y-Gwaliau south of Bala, for example.

CONSERVATION

None of the horsetails found in the mountains of Snowdonia is sufficiently rare nationally to justify statutory protection, but both Great Horsetail and Wood Horsetail are uncommon in the area. But, irrespective of whether these strange, 'prehistoric' plants are rare or not, they are a welcome addition to any habitat, and should be given every opportunity to prosper. They are, after all, among the most ancient vascular plants on Earth, and provide us with a rare glimpse as to how the vegetation on Earth looked before grasses and other flowering plants came to dominate the planet.

ADDITIONAL READING

Jermy, A. C. 1996. British Horsetails. *British Wildlife*, 8: 37-41.

Page, C. N. 1972. An interpretation of the morphology and evolution of the cone and shoot of Equisetum. *Botanical Journal of the Linnean Society*, 65: 359-397.

Page, C. N. 1988. *Ferns – their habitat in the British and Irish Landscape*. The New Naturalist. Collins, London.

Pryer, K. M., Schneider, H., Smith, A. R., Cranfield, R., Wolf, P. G., Hunt, J. S. & Sipes, D. 2001. Horsetails and ferns are a monophyletic group and the closest living relatives to seed plants. *Nature*, 409: 618-621.

MOSSES AND LIVERWORTS

TIM BLACKSTOCK and MARCUS YEO

INTRODUCTION

Mosses and liverworts, known collectively as bryophytes, include a varied assortment of plants, but they all lack the elaborate vascular systems of higher plants. Some mosses, such as *Polytrichum*, have a very simple conducting system composed of a central strand of specialist cells, and this has been used as evidence that the mosses are derived from primitive vascular plants by regressive evolution. The evolutionary origins of bryophytes are still being researched, but there is little doubt that they have a very ancient lineage. Their fossil record stretches back to at least the Devonian Period over 350 million years ago, and various of the fossil species found in Devonian strata are very similar to species that can be found living in Snowdonia today. For example, among the liverworts, *Haplomitrium hookeri* has been referred to as a 'living fossil' because of its similarity to Devonian fossils, and the thalloid liverwort *Pallavicinia lyellii* is very similar to the Devonian fossil *Pallaviciniites devonicus*. Among the mosses, the common *Mnium hornum* is close to the Permian fossil *Initia vermicularis*, which lived some 280 million years ago, and the Permian fossil *Protosphagnum* had leaves very similar to living *Sphagnum* moss. The bryophytes therefore include a number of groups that have been separated for a very long time; the ancestral forms possibly being among the earliest terrestrial plants, possibly appearing during the Silurian Period some 400 million years ago or even earlier.

In addition to their lack of vascular tissue, mosses and liverworts have similar reproductive strategies. Like all primitive plants, they produce spores rather than seeds, and these are usually released from elevated capsules. On germination the spores develop into the sexual or gametophyte generation, which produces male (antheridia) and female (archegonia) sex organs. At maturity, the antheridia produce motile sperm cells (spermatozoids), which require a film of water in which to swim to the archegonia. Fertilization takes place when the sperm cell penetrates the archegonium and fuses with the enclosed egg cell. The fertilized egg then grows and gives rise to the spore-producing generation, known as the sporophyte, which remains attached to the gametophyte and, indeed, cannot exist independently. In this respect mosses and liverworts differ from the ferns, where the sporophyte is the prominent generation. Bryophytes are also able to spread clonally through a variety of asexual mechanisms, including gemmae (often tiny clusters of specialised cells) and by wind dispersal of plant fragments.

Mosses and liverworts therefore have a number of features in common, but their differences are less easy to define because certain mosses resemble liverworts and *vice versa*. But in general liverworts differ from mosses in lacking a thickened nerve or midrib in the leaf, and some species, such as *Herbertus aduncus*, have leaves that are deeply lobed, while others, such as *Blepharostoma trichophyllum*, have narrowly segmented leaves, but no moss has these characters. Also unlike mosses, certain liverworts, known as thalloid species, such as *Conocephalum conicum,* lack a clear differentiation into stem and leaves. Finally, and less obviously, most mosses have a sporophyte capsule and stalk that persist over several months, whereas in most liverworts the final phase of sporophyte maturation is fleeting and easily overlooked.

THE HISTORY OF BRYOLOGICAL EXPLORATION IN SNOWDONIA

The bryophyte flora of Snowdonia has become particularly well known since John Jacob Dillenius, the German bryologist, visited the district in 1726. He explored the area in company with Samuel Brewer who had taken up field botany following a successful career in the English wool industry. Together they made important bryological discoveries including several species new to science (e.g. *Andreaea alpina, Bryum alpinum, Polytrichum alpinum, Racomitrum ericoides* and *Scapania ornithopodioides*). During their visit they explored Cadair Idris, Snowdon and the Glyderau, and recorded a number of characteristic upland

Bryum alpinum on Moel yr Ogof.

Breutelia chrysocoma in Cwm Idwal.

species, including *Breutelia chrysocoma, Campylopus atrovirens, Herbertus aduncus* and the uncommon *Dicranum scottianum*. Another early bryological pioneer was J. Wynne Griffith, who found the curious gemma-bearing moss *Oedipodium griffithianum* on Snowdon in the late 1700s. This also proved to be new to science and was subsequently named after Griffith. Further new records accumulated during the Victorian era, largely through the work of botanists from England, including William Wilson of Warrington, author of *Bryologia Britanica* (1855). Wilson began collecting bryophytes in North Wales in 1826, but he was especially active between 1828 and 1833, and continued making notable finds until 1867. His discoveries included the first British record of *Fissidens polyphyllus*, found close to the river near Pont Aberglaslyn, where it can still be seen today.

Up until the latter part of 19th century most of the bryological exploration in Snowdonia had been carried out in the north of the region, although the importance of the south had been hinted at in 1839 when the the lichenologist Rev. T. Salwey discovered the rare moss *Bartramidula wilsonii* at Cwm Bychan near Harlech. In due course Coed Ganllwyd, one of the best woodlands for oceanic bryophytes in Snowdonia, came to light in 1877, as a result of explorations by C. J. Wild and W. H. Pearson. By the end of the 19th century the first comprehensive list of bryophytes found in Snowdonia had been published by J. E. Griffith of Bangor in his *Flora of Anglesey and Caernarvonshire* (1895). Although his main interest was in vascular plants, Griffith did contribute bryophyte records, including the discovery in Snowdonia of *Leptodontium recurvifolium*.

While the bulk of the distinctive upland and Atlantic bryophytes of Snowdonia had been recorded by the end of the Victorian era, further important discoveries have been made in the 20th century, and the flora of the region as a whole is now well documented.

A major contributor of bryological records from Snowdonia straddling the 19th and 20th centuries was D. A. Jones, a schoolmaster from Harlech, who had an intimate knowledge of the choicest upland habitats. Most of his more important finds, including *Gymnocolea acutiloba*, *Riccia crozalsii* and *Scapania nimbosa*, were liverworts, suggesting perhaps that the group had previously been neglected in comparison to mosses. He continued to record bryophytes in Snowdonia well into the 20th century, and in 1931 organised a meeting of the British Bryological Society in Snowdonia, based at Harlech.

During the 20th century Snowdonia continued to attract many bryologists including H. N. Dixon, author of the *Student's Handbook of British Mosses* (1924), a book that has remained a valuable reference work to the present day. Although Dixon spent only short periods in Snowdonia, he was responsible for the first record of the rare *Encalypta alpina* in Cwm Dyli in 1924. In 1947 some additional bryophyte records for the north-eastern part of Snowdonia were collected and published by Albert Wilson (1947), and the British Bryological Society continued make a valuable contribution with meetings in Bangor in 1949, Beddgelert in 1970, Bangor in 1978, and Dolgellau in 1996.

In more recent times, P. M. Benoit of Abermaw (Barmouth) has followed in the footsteps of D. A. Jones, and acquired a remarkable knowledge of the flora of Meirionnydd and other parts of North Wales. Bangor has also been a centre of bryological expertise with several leading authorities based at the School of Plant Biology in the University College of North Wales, including M. O. Hill (later at the Institute of Terrestrial Ecology in Bangor), J. G. Duckett, D. A. Ratcliffe, P. W. Richards and A. J. E. Smith.

P. W. Richards became Professor of Botany at Bangor in 1949, and by the 1950s he had introduced a card-index system that brought recording in the area on to a more systematic footing. His interests included the montane and oceanic bryophytes of North Wales, and much of the bryological history of Snowdonia presented in this chapter is from his paper on the bryological exploration in North Wales (Richards, 1979). A. J. E. Smith is especially known for his publications *The Moss Flora of Britain and Ireland* (1990) and *The Liverwort Flora of Britain and Ireland* (1991), but he also maintains an interest in the local flora.

M. O. Hill was active in the field during the 1970s and 1980s, culminating in his publication in 1988 of *A Bryophyte Flora of North Wales*. This came some 93 years after Griffith's flora, and its thoroughness is testament to the importance of Hill's contribution. It is a comprehensive overview of the bryoflora of the region, with notes on distribution and ecology. Valuable contributions have also come from non-resident bryologists including H. H. Birks, H. J. B. Birks and J. A. Paton.

IMPORTANT HABITATS FOR BRYOPHYTES IN SNOWDONIA

Over North Wales as a whole, about 550 mosses and 220 liverworts (some 75% of the British flora) have been recorded, and the mountains of Snowdonia are of outstanding interest. Although temperate species are well represented, there are two other biogeographical groups that make the flora especially distinctive. The first comprises the arctic-alpine and other montane species for which Snowdonia is the main southern outpost in Britain, and the second comprises the oceanic species associated with the mild and moist Atlantic fringe of Europe.

Bryophyte distribution is to a great extent controlled by climate which, in montane regions, varies strongly with altitude. For such a small area the variation in climatic conditions within Snowdonia is surprisingly large. Over just a few kilometres, one can move from an almost frost-free maritime climate on the coast with low rainfall (870 mm on Ynys Enlli (Bardsey Island)), to sub-arctic conditions with severe winter frost and up to 3500 mm of precipitation (rain and snow) on the mountain summits.

There is no limestone in the Snowdonia heartland, but the presence of basic igneous bedrock greatly enhances bryophyte diversity. The scarcity of such rocks in the hills to the south of Cadair Idris appears to be the major reason for their floristic poverty.

THE ALPINE ZONE

Many of the montane species in Snowdonia are at the southern edge of their British range. A list of montane mosses and liverworts is given in Table 1. The species included are most frequent above 600 m, but many descend to lower altitudes. Reasonable claims for the addition of other species could be made and the selection of representatives for such a group is somewhat arbitrary.

Table 1. Mosses and liverworts that are mainly restricted to montane habitats recorded from Snowdonia

Montane mosses	Montane liverworts
Amblyodon dealbatus	Anthelia julacea
Amphidium lapponicum	Anthelia juratzkana
Andreaea alpina	Diplophyllum taxifolium
Andreaea mutabilis	Eremonotus myriocarpus
Anoectangium warburgii	Gymnomitrion concinnatum
Arctoa fulvella	Gymnomitrion corallioides
Bryum mildeanum	Gymnomitrion crenulatum
Bryum muehlenbeckii	Gymnomitrion obtusum
Bryum weigelii	Herbertus stramineus
Conostomum tetragonum	Jungermannia borealis

Table 1 continued	
Montane mosses	**Montane liverworts**
Dicranoweisia crispula	Marsupella adusta
Ditrichum zonatum	Marsupella alpina
Encalypta alpina	Marsupella stableri
Grimmia elongata	Marsupella sphacelata
Hypnum hamulosum	Scapania aequiloba
Isopterygiopsis muelleranum	Scapania calcicola
Kiaeria blyttii	Scapania gymnostomophila
Kiaeria falcata	Scapania nimbosa
Mnium thomsonii	Scapania ornithopodioides
Myurella julacea	Scapania paludosa
Orthothecium rufescens	Scapania uliginosa
Philonotis seriata	
Philonotis tomentella	
Plagiothecium denticulatum var. obtusifolium	
Plagiothecium platyphyllum	
Pohlia ludwigii	
Pseudoleskeella catenulata	
Pterigynandrum filiforme	
Racomitrium macounii	
Schistidium trichodon	
Tetraplodon augustatus	

All the species in Table 1 are also found in the Scottish mountains and many, though not all, have been recorded from northern England. Examples of species known from the North Wales mountains but not from England include *Bryum muehlenbeckii, Isopterygiopsis muelleranum, Philonotis seriata, Jungermannia borealis* and *Scapania gymnostomophila*. We estimate that out of 146 montane species in Britain, 100% are recorded from Scotland, some 38% from England and about 36% from Wales. These figures are comparable to equivalent data for montane vascular plants; out of 118 species of flowering plants and ferns (fewer montane species than the bryophytes), 95% occur in Scotland, 54% in England and 38% in Wales.

Many of the montane mosses and liverworts are represented in Snowdonia by very small populations. They are at the climatic margins of their distribution in Britain, and are often confined to microhabitats where competition from more abundant taxa is low. Many of the high-altitude species were probably more frequent and widely distributed during the early postglacial period.

ROCK HABITATS

Among the most noticeable features of the Snowdonian landscape are the prominent crags and cliffs, smaller rock outcrops and scree slopes. These support a large and diverse bryophyte flora, with species composition determined by variables such as rock type, aspect, altitude and exposure.

Acidic rocks such as rhyolite predominate over most of Snowdonia, and above the main limit of agricultural enclosure at about 300-400 m, many outcrops and rock slabs are covered predominantly by lichens. They are often accompanied, though, by various submontane bryophytes, particularly from the moss genus *Racomitrium*. The hoary tufts of *R. lanuginosum*, for example, are a familiar sight on acidic boulders and outcrops, and equally common are the matt green patches of *R. affine*, *R. aquaticum* and *R. fasciculare*, the second typically occurring on sloping rock faces where there is occasional flushing. Also in such places are the dark green, almost black, tufts of *Campylopus atrovirens*, occasionally accompanied by its scarcer relative *C. schwarzii*. Black or dark red patches of *Andreaea rothii* and *A. rupestris* are another distinctive feature of acidic outcrops and boulders in the unenclosed uplands, and more locally may be found the curious compact tufts of the liverworts *Gymnomitrion crenulatum* and *G. obtusum* with their wiry stems and tiny closely overlapping leaves.

Gymnomitrion obtusum with *Diplophyllum albicans* on Tryfan.

At higher altitudes the flora of acidic rocks may include montane species such as *Dicranoweisia crispula*, *Kiaeria blyttii*, *Marsupella adusta* and *M. stableri*. Most of these species are restricted to sheltered microhabitats within the boulder screes that form below many of the cliffs and crags in the main Snowdonian mountains. Occasionally in the patches of peaty soil between the boulders of these high altitude screes can be found the moss *Oedipodium griffithianum*.

Oedipodium griffithianum on Moel yr Ogof.

The bryophyte flora of base-rich rocks such as pumice tuff and dolerite is very different to that of acidic substrates. At low to moderate altitudes the drier rock faces support widespread species such as *Hypnum lacunosum*, *Neckera crispa*, *Pterogonium gracile* and *Frullania tamarisci*. On higher ground, the

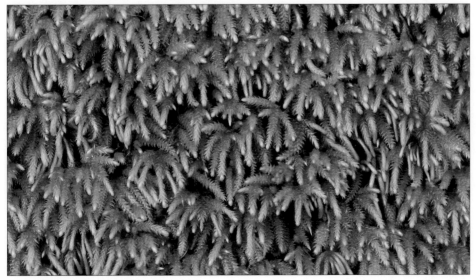

Pterogonium gracile in Nantgwynant.

greyish-green tufts of *Grimmia torquata* are typical of relatively dry calcareous rocks, and at one locality in northern Snowdonia, *Tortula princeps* occurs in a similar habitat at its only Welsh site. Flushed rock faces are often distinguished by olive-green patches of *Blindia acuta*, with associated species such as *Amphidium mougeottii* and *Racomitrium ellipticum*, and the minute leafy liverwort *Hygrobiella laxifolia*.

Sheltered base-rich rock crevices and ledges support a remarkable assemblage of species, including many upland rarities. Such microhabitats are usually dominated by acrocarpous mosses, frequently including compact bright green tufts of *Anoectangium aestivum*, dark green tufts of *Amphidium mougeottii* and coarse tufts of *Tortella tortuosa*. Other characteristic species found growing over or within the clumps of moss include *Fissidens cristatus*, *Isopterygium pulchellum*, *Pohlia cruda*, *Plagiochila porelloides* and *Preissia quadrata*. Rarer species, restricted to a few localities in Snowdonia, include *Amphidium lapponicum*, *Anoectangium warburgii*, *Hypnum hamulosum*, *Isopterygiopsis muelleranum*, *Mnium thomsonii*, *Orthothecium rufescens*, *Herbertus stramineus*, *Jungermannia borealis*, *Leiocolea heterocolpos*, *Scapania aequiloba* and *S. gymnostomophila*. A notable rarity is the minute moss *Seligeria brevifolia*, first discovered on Snowdon in 1978 by M. O. Hill and now known from three sites on the mountain, in each case growing in small patches beneath overhangs. Outside Snowdonia it is known only from one or two localities in Scotland and one in England.

Polytrichum piliferum in Nantgwynant.

The richest sites for bryophytes are found at the bases of high cliffs where there is a mixture of acid and calcareous rock types. Among the best are Clogwyn Du'r Arddu, Clogwyn y Garnedd, Cwm Glas Bach and Cwm Glas Mawr (in the Snowdon range), Cwm Bochlwyd and Cwm Idwal (in the Glyderau), Ysgolion Duon (below Carnedd Dafydd), Moel yr Ogof (to the southwest of Snowdon) and the crags in the vicinity of Llyn Gafr and Llyn y Gadair (in the Cadair Idris range). Several of these localities are within National Nature Reserves and all are in protected conservation sites.

At mostly low altitudes, old disused slate quarries, such as parts of the Penrhyn and Dinorwig quarries, often have extensive carpets of mosses such as *Polytrichum piliferum, Racomitrium lanuginosum* and *R. fasciculare*. There is also a highly specialised small group of species associated with copper and lead ores. Several disused lead mines are found in Coed Gwydir, and for most plants the waste material is highly toxic, but *Ditrichum plumbicola* is restricted to such environments and has been recorded at some 20 Gwydir localities. *Grimmia atrata* is another moss tolerant of heavy metals on old mine workings, but it also occurs on undisturbed rock faces at altitudes from about 200-900 m.

The mosses *Racomitrium lanuginosum* and *R. fasciculare* in Penrhyn Slate Quarry.

WOODLAND

Despite extensive deforestation over several thousand years of human occupation, patches of woodland still persist in many areas of Snowdonia, especially on rocky hill ground and in deep ravines; locally, as in parts of Meirionnydd, there are more extensive tracts of semi-natural woodland. These woodlands, typically dominated by oak and to a lesser extent birch, often hold a rich assemblage of bryophytes, including several oceanic Atlantic species. Although this element of the flora is not as well developed here as in north-west Scotland or the south-west of Ireland, it still holds a fascination for bryologists.

Hylocomium splendens in marshes near Tregarth.

Leucobryum glaucum in Llwyn y Coed near Llanberis.

In humid upland woods (mostly found below 300 m altitude) the ground is typically carpeted with luxuriant mats of bryophytes, in which the bulk of the cover is provided by robust mosses such as *Dicranum majus, Hylocomium splendens, Leucobryum glaucum, Rhytidiadelphus loreus* and *Thuidium tamariscinum.*

Thuidium tamariscinum in Coed Victoria.

More locally there may be deep mats of liverworts such as *Adelanthus decipiens, Bazzania trilobata, Lepidozia cupressina* and *Plagiochila spinulosa*, together with *Dicranodontium denudatum, Hylocomium umbratum* and other mosses. The role of bryophytes in relation to nutrient dynamics within such woodlands has been studied by Rieley *et al.* (1979). The chemistry of woodland soil can

Lepidozia cupressina in Coed y Rhygen.

exert a marked influence on the bryophyte flora. On base-rich ground the acid-loving species (calcifuges) are replaced to some extent by mildly lime-loving species (calcicoles), including *Cirriphyllum piliferum, Eurhynchium striatum, Hylocomium brevirostre* and *Plagiochila asplenioides*, while species such as *Calliergon cordifolium, Rhizomnium punctatum* and *Trichocolea tomentella* predominate in the wetter spots.

The characteristic woodland-floor species often extend over rocks and boulders, and may be joined by rarities such as *Plagiochila atlantica*. Close scrutiny of damp, sheltered rock faces, especially in ravines, can reveal various species of minute leafy liverworts, including *Colura calyptrifolia, Drepanolejeunea hamatifolia, Harpalejeunea ovata, Plagiochila corniculata* and *Radula voluta*, which are restricted to the areas of high rainfall in the west of Britain. Very locally the slender green stems of *Sematophyllum demissum* can be found. This moss is now known from only about seven localities in Snowdonia but is almost invariably found in small quantity. Unlike most other Atlantic species, it does not occur in the oak woods of western Scotland, but is known from western Ireland. Its relative *S. micans* was presumed extinct in the Cwmnantcol ravine, just south of Harlech (its only known locality in Wales), until it was refound there in 1999. Rocks in and by woodland streams usually support specialised semi-aquatic bryophytes such as the dark green *Jubula hutchinsiae*, an oceanic liverwort usually seen under dripping overhanging rocks and boulders. In the spray zone of some ravine sites, the very scarce moss *Campylopus setifolius* finds a congenial microhabitat. It is one of several strongly oceanic species that extend above the modern woodland zone to sheltered rocky ground at somewhat higher altitudes. Another of the great rarities of the region, *Rhytidiadelphus subpinnatus*, has been recorded from two ravine woodlands in Meirionnydd, but

Sematophyllum demissum in Coed Ganllwyd.

Campylopus setifolius in Coed Ganllwyd.

has been seen recently at only one of these localities, where it was rediscovered in the 1980s after a gap of over 100 years.

Tree bases are typically clothed in *Isothecium myosuroides*, while trunks and branches are generally dappled red and green with mixtures of *Hypnum andoi*, *Frullania tamarisci*, *Metzgeria temperata* and *Plagiochila punctata*. The minute yellow-green stems of *Microlejeunea ulicina* can often be found growing among these larger species, and at a couple of sites, *Leptoscyphus cuneifolius* occurs in similar situations. Rotting wood also has a characteristic assemblage of liverworts. Logs may be conspicuously coloured, even at a distance, by reddish sheets of *Nowellia curvifolia*. Rarer species include *Cephalozia catenulata*, *Tritomaria exsecta* and the small thalloid liverwort *Riccardia palmata*.

Nowellia curvifolia in Capel Curig woods.

Plantations of non-native conifers are widespread in Snowdonia, but in contrast to semi-natural woodlands they generally have a very limited flora composed mostly of a few common species of acid habitats. The only notable species is the moss *Plagiothecium curvifolium*, which in North Wales is restricted to banks and logs beneath conifers; it appears to be a recent introduction to the area.

The richest woodlands in Snowdonia are in the valleys of the Harlech dome that drain the Rhinogau, and several National Nature Reserves (NNRs) and other conservation sites have been established to protect their bryophytes. Especially well known are Coed Ganllwyd NNR and Coed y Rhygen NNR. Further north, there are good sites in the foothills of Snowdon and the Glyderau, and in Dyffryn Conwy.

As mentioned earlier, many of the distinctive Atlantic woodland species ascend to some degree into the higher unwooded hill country, especially in sheltered situations. There is, for example, a good oceanic flora on the steep wet rock faces of the Ogwen Falls at an altitude of 250 m in the upper reaches of Nant Ffrancon. It is likely that many such sites formerly had a woodland or scrub cover, but although the trees are now long gone, these slopes are still sufficiently mild and moist to permit the patchy survival of certain oceanic mosses and liverworts.

GRASSLAND AND HEATH

Above the mountain wall, the extensive sheep-grazed *Agrostis-Festuca* and *Nardus* grasslands are generally poor in bryophytes. For the most part, only a few relatively robust mosses are frequent and abundant, including *Hylocomium splendens*, *Pleurozium schreberi*, large *Polytrichum* species, *Pseudoscleropodium purum* and *Rhytidiadelphus squarrosus*. Patches of calcareous grassland hold a different range of species, but these are also mostly widespread plants, such as *Ctenidium molluscum*. One exception is the distinctive moss *Rhytidium rugosum*, which in North Wales is restricted to high-altitude base-rich grasslands, as on Moel Hebog and in Cwm Glas Mawr.

Where grazing pressure is less intense, grasslands are replaced by dry heaths in which dwarf shrubs such as *Calluna vulgaris*, *Erica cinerea* and *Vaccinium myrtillus* are dominant. Such communities may support a luxurious carpet of bryophytes, but species diversity is typically very limited and the bulk of the cover is provided by the large mosses characteristic of acid grasslands, supplemented by a few additional species such as *Plagiothecium undulatum*. Patches of bare soil may be colonised by mosses such as *Campylopus paradoxus* and *Pohlia nutans*. Wetter heaths on waterlogged soils often have patches of *Sphagnum compactum* and *S. tenellum*, together with other species of damp peat, such as the leafy liverwort *Gymnocolea inflata*.

The richest forms of heath vegetation are found on steep rocky slopes with a north or easterly aspect in the high rainfall zone of the western hills. As well as the more typical mosses, these humid heaths have a well-developed bryophyte layer that often includes a *Sphagnum* component, with *Sphagnum capillifolium*,

Pseudoscleropodium purum in Cwm Idwal.

Sphagnum spp. near Capel Curig.

S. palustre and *S. quinquefarium* among other species. Also present may be several Western and Atlantic taxa such as liverworts like *Anastrepta orcadensis, Bazzania tricrenata, Herbertus aduncus* subsp. *hutchinsiae* and *Lepidozia pearsonii*. The rare *Gymnocolea acutiloba* is recorded at two localities. Such communities are much more extensive and better developed in the north-west highlands of Scotland, but the Snowdonian examples demonstrate the wide distribution of strongly oceanic elements in the upland flora of western Britain. Particularly good examples are found on the heather-clad hillsides of the Rhinogau.

Towards the higher summits, frost-disturbed rocky soil banks have a good bryophyte cover with common species such as the moss *Oligotrichum hercynicum* and the liverworts *Diplophyllum albicans* and *Nardia scalaris*. On the higher summit plateaux, exposed montane grass-heaths characteristically have extensive cover of wind-blasted *Racomitrium lanuginosum* and few associates. Although there are one or two small areas in Snowdonia where snow lies relatively late into the year, there are no true snow-beds, and the specialist mosses and liverworts associated with such habitats in the Scottish Highlands are not found in Wales.

The abundance of livestock in Snowdonia is responsible for a group of uncommon but distinctive mosses that grow on dung – the so-called coprophilous (dung-loving) species. *Splachnum ampullaceum* and *S. sphaericum* are found occasionally throughout the region, the former usually on cow dung in wetlands, and the latter on sheep dung on moorlands. The closely related *Tetraplodon mnioides* is a plant of sheep carcasses in the Snowdonian uplands. All these species are characterised by large and brightly coloured sporophytes, and a reliance on insects for spore dispersal, rather than simply by wind as is usual in mosses.

MIRES AND FRESHWATER HABITATS

Blanket bog is rather local over much of Snowdonia, but can be extensive where the topography has allowed peat to accumulate. In the drier forms the only *Sphagnum* species may be *S. capillifolium,* occurring as scattered hummocks, together with other common mosses such as *Hypnum jutlandicum* and *Pleurozium schreberi*. *Sphagnum* species become more abundant in wetter areas where there may be extensive lawns of *S. capillifolium, S. papillosum* and *S. subnitens*. *Sphagnum cuspidatum* grows submerged in bog pools. Scattered stems of the leafy liverwort *Odontoschisma sphagni* are often found within *Sphagnum* hummocks, occasionally joined by other leafy liverworts such as *Cladopodiella fluitans, Kurzia pauciflora* and *Mylia anomala*. Eroding peat banks sometimes have a distinctive assemblage of liverworts, including *Calypogeia neesiana, Kurzia trichoclados* and *Odontoschisma denudatum*. Raised bogs are very uncommon in the district but there are examples to the south of Llyn Trawsfynydd and at Cors Arthog (Arthog Bog) on the south side of the Mawddach Estuary, and these hold a number of rare bryophytes, including *Sphagnum imbricatum* subsp. *austinii* (which was formerly a major peat-builder) and the liverwort *Pallavicinia lyellii*.

Acid flush vegetation is widespread at moderate to high altitudes, and typically contains common species such as *Calliergon stramineum, Polytrichum commune, Sphagnum auriculatum* and *S. recurvum*. More base-rich stands are dominated by 'brown mosses' such as *Calliergon cuspidatum, Cratoneuron commutatum, Drepanocladus revolvens* and *Scorpidium scorpioides*, together with *Aulocomnium palustre, Bryum pseudotriquetrum* and *Fissidens adianthoides*. The base-tolerant sphagna, *Sphagnum contortum, S. teres* and *S. warnstorfii,* occur more locally. Springs and seepages may contain swollen growths of *Bryum pseudotriquetrum, Philonotis fontana, Scapania undulata* and other bryophytes. At higher altitudes the flora is occasionally enriched by rarities such as *Bryum weigelii, Philonotis seriata* and *Scapania uliginosa* with, in a few localities, the liverwort *Anthelia julacea* forming extensive silver-green cushions.

Fast-flowing mountain streams often have a profusion of bryophytes, and several species, including *Fontinalis squamosa, Hygrohypnum ochraceum, Hyocomium armoricum, Marsupella emarginata, Nardia compressa* and *Scapania undulata*, may form extensive growths. Substrates that are mildly base-rich are indicated by the presence of species such as *Hygrohypnum eugyrium* and *Thamnobryum alopecurum. Bryum muehlenbeckii* is a rare species of streamside rocks in two localities in Snowdonia; otherwise it is confined in Britain to the north Scottish highlands. Certain streamside species such as the thalloid liverwort *Conocephalum conicum* and also occasionally *Lunularia cruciata,* are more frequent at low altitudes.

Polytrichum commune in Cwm Idwal.

Scorpidium scorpioides on Moel Hebog.

Aulocomnium palustre with *Sphagnum* in Tregarth woods.

Philonotis fontana on Elidir Fach.

Anthelia julacea near Llyn y Cŵn on the Glyderau.

Lunularia cruciata in Afon y Llan (Coed Cochwillan).

Very few bryophytes are found constantly submerged in deep-water lakes, but lake and reservoir margins can have a diverse flora including scarce species, such as *Haplomitrium hookeri*, known from Llyn Idwal and two sites in the eastern Carneddau. Other characteristic species of such habitats include ephemeral mosses such as *Ephemerum serratum*, *Pohlia bulbifera* and *Pseudephemerum nitidum*.

ADJACENT COASTLANDS AND THE MANAGED LOWLANDS

Cliff tops, rock faces and south-facing sunny banks on the coast have a varied flora which includes several Mediterranean species at or near the northern limit of their range; *Fissidens algarvicus*, *Tortula cuneifolia*, *Fossombronia maritima* and *F. husnotii* are examples. Sand dunes are another important habitat in north-west Wales and hold a large specialist flora. The dune system at Aberffraw in Ynys Môn (Anglesey) is particularly relevant here as it has certain species such as *Catoscopium nigritum* and *Meesia uliginosa*, which occur in maritime dune slacks and in upland habitats in northern Britain (but not, oddly, in the case of these two species, in Snowdonia).

Further inland, semi-natural habitats are confined to the farmed lowlands where improved sheep and cattle pastures prevail. There are some interesting heaths and fens with a good range of mosses and liverworts, but the major attractions for the bryologist are the rocky hill woodlands and deep ravines which have escaped the most prolonged and destructive human activities. Overall, the lowlands now have a sparse cover of native habitats, often represented by small and fragmented patches, contrasting markedly with the situation in the uplands (above the upper level of enclosure) where there is extensive semi-natural land cover.

OCEANIC SPECIES

The Atlantic or oceanic element is represented by a number of sub-groups (Hill & Preston, 1998). Summary information on their altitudinal range in north-west Wales is given in Table 2. This covers elements that are widespread in the Mediterranean region and extend to the Atlantic seaboard, and others that are confined to southern, temperate and boreal zones on the Atlantic fringe of Europe; hyperoceanic species have a markedly western distribution. The Mediterranean and southern oceanic sub-groups are very well represented, with a number of species limited to the mild coastal zone and mostly found below 300 m. The hyperoceanic southern-temperate sub-group is especially associated with humid oak-birch woodlands and ravines. The temperate and boreal Atlantic sub-groups become increasingly prevalent at higher elevations and include a number of taxa confined or mostly restricted to upland habitats. In short, the bryophyte flora of Snowdonia is made distinctive by the suffusion of strongly oceanic species distributed along the coastal-lowland-montane gradient.

CHANGES IN THE BRYOPHYTE FLORA AND THE INFLUENCE OF MAN

In broad terms, it seems reasonable to conclude that the postglacial montane flora has become diminished and restricted as the climate ameliorated. Except for the coastal fringe and the mountain summit zone, much of Snowdonia was covered by forest until it began to be cleared some 5000 years ago by early human settlers. The vegetation cover today has been altered profoundly. In the lowlands productive sheep and cattle pastures predominate, and there is very little arable land. The hill and mountain country consists primarily of grassland sheep-walk with large areas of heath and rock habitats; woodlands are almost completely absent above 300 m. The bryophyte flora has had to adapt to major perturbations from a long history of human occupation and land use. Some species that are now present only in small isolated populations may have their dispersal capacity limited by lack of sporophyte development.

Assessment of changes in the local distribution and frequency of bryophyte taxa is greatly hampered by the paucity of historical data. Separation of the relative influence of natural causes and human impacts is often difficult. Even careful scientific investigation may be inconclusive. For example, a comparative study of the effects of land use on Atlantic bryophyte assemblages in four Snowdonia woodlands revealed no simple relationship between species composition and disturbance history (Edwards, 1986). The impact of temporary canopy removal appeared to be at least partly dependent on local microhabitat conditions.

Table 2. Oceanic elements in the Snowdonia bryophyte flora. Altitudinal range in north-west Wales is indicated by: C = Coastal, L = Lowland (below 300m), U = Upland (higher than 300 m). Biogeographic groups follow Hill and Preston (1998)

1. Mediterranean-Atlantic Group

Species	C	L	U
Anthoceros punctatus		L	
Bryum canariense		L	
Bryum donianum	C	L	
Bryum torquescens		L	
Cinclidotus mucronatus		L	
Epipterygium tozeri		L	
Fissidens algarvicus	C		
Fissidens limbatus	C	L	
Fissidens serrulatus		L	
Fossombronia angulosa	C		
Fossombronia husnotii	C		
Funaria attenuata		L	U
Grimmia britannica		L	U
Grimmia lisae		L	U
Gymnostomum viridulum	C		
Leicolea turbinata	C	L	
Leptodon smithii		L	
Petalophyllum ralfsii	C		
Phaeoceros laevis ssp. laevis		L	
Philonotis rigida	C	L	
Pottia commutata	C		
Pottia crinita	C		
Pottia recta	C	L	
Pottia wilsonii	C	L	
Riccia crozalsii	C		
Riccia nigrella	C		
Scleropodium tourettii	C	L	
Scorpiurium circinatum		L	
Southbya tophacea	C		
Targionia hypophylla	C	L	
Tortella nitida		L	
Tortula canescens	C		
Tortula cuneifolia	C		

2. Oceanic Southern-temperate Group

Species	C	L	U
Campylopus polytrichoides	C		
Fissidens curnovii		L	
Fissidens monguillonii		L	
Fossombronia maritima	C		
Glyphomitrium daviesii	C	L	U
Lophocolea fragrans		L	
Marchesinia mackaii		L	
Porella obtusata	C	L	
Porella pinnata		L	U
Ptychomitrium polyphyllum		L	U
Saccogyna viticulosa	C	L	U
Ulota calvescens		L	U

3. Hyperoceanic Southern-temperate group

Species	C	L	U
Adelanthus decipiens		L	
Aphanolejeunea microscopica		L	
Cololejeunea minutissima	C	L	
Colura calyptrifolia		L	
Dicranum scottianum		L	
Drepanolejeunea hamatifolia		L	
Fissidens polyphyllus		L	
Frullania microphylla	C	L	
Frullania teneriffae	C	L	U
Harpalejeunea ovata		L	U
Jubula hutchinsiae		L	
Lejeunea lamacerina		L	U
Lepidozia cupressina		L	
Leptoscyphus cuneifolius		L	
Metzgeria leptoneura		L	U
Plagiochila exigua		L	U
Plagiochila killarniensis		L	
Plagiochila punctata		L	U
Radula aquilegia		L	U
Scapania gracilis		L	U

4. Oceanic Temperate group

Species	C	L	U
Campylopus brevipilus	C	L	U
Dounia ovata		L	U
Fissidens celticus		L	
Hedwigia integrifolia		L	U
Hyocomium armoricum		L	
Isothecium holtii		L	
Jungermannia paroica		L	U
Leptodontium flexifolium		L	U
Orthotrichum pulchellum		L	
Orthotrichum sprucei		L	
Sematophyllum demissum		L	
Sematophyllum micans		L	

5. Hyperoceanic Temperate group

Species	C	L	U
Andreaea megistospora		L	U
Bartramidula wilsonii		L	
Breutelia chrysocoma		L	U
Bryum riparium		L	U
Campylopus atrovirens		L	U
Campylopus setifolius		L	U
Isothecium myosuroides var. brachythecioides			U
Lepidozia pearsonii		L	U
Plagiochila atlantica		L	
Radula voluta		L	
Sphagnum imbricatum ssp. austinii		L	
Sphagnum strictum			U

6. Oceanic Boreo-temperate group

Species	C	L	U
Schistidium maritimum	C		
Ulota phyllantha	C	L	

7. Oceanic Boreal-montane group

Species	C	L	U
Andreaea alpina			U
Andreaea mutabilis			U
Anoectangium warburgii			U
Gymnomitrion crenulatum		L	U
Herbertus aduncus ssp. hutchinsiae		L	U
Herbertus stramineus			U
Leptodontium recurvifolium		L	U
Oedipodium griffithianum			U
Racomitrium ellipticum			U
Rhabdoweisia crenulata		L	U
Scapania nimbosa			U
Scapania ornithopodioides			U

Among the species that appear to have become recently extinct in Snowdonia there are several montane and oceanic bryophytes that have not been seen since the 19th century or the early part of the 20th century (Table 3). Although losses of high-altitude species might be expected, the probable regional extinction of ten or so montane mosses and liverworts over the last 150 years is remarkable when compared to the persistence of upland vascular plants during the same period. Some reductions may have been exacerbated by collecting, but bryologists cannot be held responsible for a decline on this scale. Climate change and habitat alteration are both likely to have played significant roles.

Table 3. Bryophytes that have appear to have recently become extinct in Snowdonia

Mosses	Date of last record		Date of last record
Bartramidula wilsonii	1931	Pterigynandrum filiforme	1928
Bryum mildeanum	1892	Schistidium trichodon	1907
Cinclidotus mucronatus	1927	Tetraplodon angustatus	1899
Conostomum tetragonum	1919	**Liverworts**	
Encalypta alpina	1931	Diplophyllum taxifolium	1844
Glyphomitrium daviesii	1917	Gymnomitrion corallioides	1912
Myurella julacea	1912	Scapania nimbosa	1909

Pollution from atmospheric sources seems unlikely in this western district remote from industrial centres, but conservationists are increasingly concerned about the impacts of nitrogen deposition and other pollutants. Enhanced nitrogen levels may have adverse effects on *Sphagnum* and possibly *Racomitrium lanuginosum* and other species; but it may also have a wider influence on the vegetation through increased nutrient loading. Again it can also be difficult to differentiate between the parts played by pollution and land use.

Prospects for the future survival of the distinctive bryoflora will be dependent on our ability to manage habitat resources wisely. As well as curtailing the impact of destructive agencies, there is still a requirement to improve our understanding of the relationships between bryophytes and their environment.

The importance of maintaining the distinctive bryoflora of Snowdonia is emphasised by the fact that many of the species are very rare or scarce in Britain, and several, including *Amblystegium saxatile, Bartramidula wilsonii, Fissidens serrulatus, Rhytidiadelphus subpinnatus, Sematophyllum demissum, Cephaloziella massalongi, C. nicholsonii, Gymnocolea acutiloba, Pallavicinia lyellii* and *Riccia nigrella* have been recently included in a Red Data Book of Mosses and Liverworts (Church *et al.,* 2001). The rare moss *Drepanocladus vernicosus* is also now protected under both British law (Wildlife and Countryside Act) and European legislation (EC Habitats and Species Directive).

ADDITIONAL READING

Edwards, M. E. 1986. Disturbance histories of four Snowdonian woodlands and their relation to atlantic bryophyte distribution. *Biological Conservation*, 37: 301-320.

Church, J. M., Hodgetts, N. G., Preston, C. D. & Stewart, N. F. (eds) 2001. *British Red Data Books. Mosses and Liverworts*. Joint Nature Conservation Committee, Peterborough.

Hill, M. O. 1988. A bryophyte flora of North Wales. *Journal of Bryology*, 15: 377-491.

Hill, M. O. & Preston, C. D. 1998. The geographical relationships of British and Irish bryophytes. *Journal of Bryology*, 20: 127-226.

Hodgetts, N. 1993. Atlantic bryophytes on the western seaboard. *British Wildlife*, 4: 287-295.

Richards, P. W. 1979. A note on the bryological exporation of north Wales. In: G. C. S. Clarke & J. G. Duckett (eds.), *Bryophyte Systematics*, pp.1-9. Academic Press, London and New York.

Rieley, J. O., Richards, P. W. & Babbington, A. D. L. 1979. The ecological role of bryophytes in a North Wales woodland. *Journal of Ecology*, 67: 497-527.

Smith, A. J. E. 1990. *The Moss Flora of Britain and Ireland*. Cambridge University Press, Cambridge.

Smith, A. J. E. 1991. *The Liverworts of Britain and Ireland*. Cambridge University Press, Cambridge.

Wilson, A. 1947. The flora of a portion of north-east Caernarvonshire. *Northwestern Naturalist*, 22: 191-211.

FUNGI

CHARLES ARON

INTRODUCTION

The fungi comprise a very diverse group of plant-like organisms, but unlike green plants, they are made up of minute filaments known as hyphae. These are usually composed of a protein known as chitin, which is also the main constituent found in the outer skeletons of insects and other invertebrates. This contrasts with the cellulose cell walls of green plants, and shows that fungi are as distinct from plants as they are from animals and occupy their own separate Kingdom.

When hyphae mass together they form a structure known as a mycelium, which is a network of fungal threads that remains immersed in the substrate, and absorbs nutrients from it. The mycelium is in fact the fungal colony and it may be very extensive. Large areas of typical woodland, for example, may be permeated by the mycelia of many different species, some of which may persist for many years. When conditions of temperature and humidity are right, which usually occurs in the autumn for many species, the mycelium produces the fruiting bodies that we see in the field. Mushrooms, brackets and cups are some of the more familiar forms, but there are others that are less well known, and even some, the truffles (hypogeous fungi), that produce fruiting bodies underground.

Fungi, unlike green plants, do not contain chlorophyll, and so are unable to use sunlight to synthesize organic molecules. Instead, they depend on other organisms, plant or animal, for their sustenance. By far the largest group, the saprophytes, contains the species that are responsible (along with bacteria) for the breakdown and recycling of dead organic matter, and are absolutely essential to the continuance of life on Earth. A second group, the parasites, contains species that derive their sustenance from the living tissues of plants, animals or even other fungi. Many pathogens, especially of those of plants, fall into this category, but they are usually small or microscopic and, with few exceptions are outside the scope of this book. The third group, the symbiotic fungi, are associated only with living plants, but with a degree of mutual dependence. They combine with algae to form lichens, and with the roots of flowering plants as mycorrhizae (meaning fungus root) in a way that benefits both. In the case of mycorrhizae their mycelia form a sheath that surrounds or actually penetrates root hairs, enabling them to absorb nutrients directly from the plant. For its part, the plant gains an enhanced ability to extract water and minerals from the soil. The importance of these associations to the survival of higher plants is being increasingly recognized, and we now know that many forest trees including common species such as oak, birch and pine, require them for proper growth.

The earliest fossils to be definitely identified as fungi were alive some 500 million years ago in the Ordovician Period, but fossil filaments resembling fungi have been found in rocks up to 900 million years old. In fact, it is now thought that fungi may have evolved independently from bacteria-like organisms living as much as two billion years ago, and therefore may predate the evolution of all but the most primitive of single-celled plants. All the major groups of fungi we find living today seem to have been in existence since at least the end of the Carboniferous Period, some 300 million years ago. Even mycorrhizal associations appear to have developed very early, with fossils of some of the first land plants showing traces of associated fungi from a primitive group known as Oomycetes.

Whatever the shape of a fungal fruiting body, its function is always the same – to manufacture and then disperse the minute reproductive cells known as spores. Classification is based on their reproductive anatomy, with the larger and more familiar fungi divided into two main groups, the basidiomycetes (spore droppers) and the ascomycetes (spore shooters).

Basidiomycetes include the familiar mushrooms and toadstools (agarics) and the bracket fungi. Their spores develop on club-shaped cells known as basidia, which line the outer surfaces of gills, folds, pores, or even spines, forming a fertile layer known as the hymenium. The most common arrangement is for this layer to occupy the underside of the 'cap', and for spores to be shed downwards, but the gasteroid (stomach fungi) such as puffballs and earthstars have an internal hymenium and release their spores through a pore or rupture in the outer layer (peridium) of the fruit body.The ascomycetes (spore shooters) include the cup fungi and morels. They differ from the basidiomycetes in producing spores within microscopic torpedo-shaped structures known as asci, which can 'fire' spores some distance away from the fungus. The asci are arranged either on the outer surface of the fungus, as in morels, helvellas and earth tongues, or as a dense palisade on the inner lining of cup fungus such as the Orange Peel Fungus (*Aleuria aurantia*).

Sexual reproduction in fungi can be complex, but in the more familiar or 'higher' fungi, the life cycle starts with the germination of spores to form primary hyphae. Then, when different but compatible mating strains come together, fusion takes place and secondary hyphae are produced. It is these that give rise to the spore-producing fruiting bodies. There are a number of variations on this theme however, and there is even one group, the Fungi Imperfecti, in which the sexual phase appears to be completely lacking.

THE HISTORY OF FUNGUS EXPLORATION IN SNOWDONIA

Among the early collectors of fungi to reach the fringes of Snowdonia were the Rev. Miles Berkeley, parson of Apethorpe in Northamptonshire, and Charles Broome from Elmhurst near Bath. In Victorian times these amateur mycologists did much pioneering work on identifying and recording fungi in Britain and overseas, and described many new species. They regularly visited Coed Coch, near Abergele, from where they made forays to the rich oak and beech woods of Dyffryn Conwy, and made a number of interesting discoveries. In the 1860s for example, they found the rare hedgehog fungi *Phellodon confluens* and *P. melaleucus* in Coed Dolgarrog – the latter has recently been recorded near Betws-y-Coed. The Thin Nut Truffle (*Hymenogaster tener*), was recorded at about the same time from nearby Trefriw, perhaps the only Welsh record we have.

Much of our knowledge of Snowdonian fungi has come from the periodic autumn and spring forays of the British Mycological Society (BMS). In September 1924, they explored the area around Betws-y-Coed and made many finds that have yet to be repeated or confirmed for north-west Wales. At Rhaeadr Ewynnol (Swallow Falls), two new hedgehog fungi, *Hydnellum ferrugineum* and *Phellodon niger,* were recorded, but neither has been re-found, and since they were the only Welsh records, it may be that they are now extinct in Wales. The group also encountered the rare cup fungi *Aleuria luteonitens* and *Ramsbottomia lamprosporoides* at Rhaiadr Ewynnol, and the Purple-black Earth Tongue (*Theumenidium atropurpureum*) at Fairy Glen – only the third 20th century record for Wales. Other interesting finds were *Cortinarius largus*, *C. phoeniceus* and *Inocybe godeyi*.

Following a dry and rather lean year for fungi in 1949, the September 1950 BMS foray, based at Bangor, was productive, with many new discoveries. The foray was perhaps most notable, though, for the truffle discoveries of Professor Lilian Hawker. She was a mycologist based at Kew, and her painstaking excavations under both deciduous and coniferous trees brought to light at least a dozen species. Most of the sites visited were in the lowland areas close to Bangor, but there were also visits to Betws-y-Coed, Cwm Idwal and the area around Llyn Ogwen. Searches in the Betws-y-Coed area yielded three truffles, *Hymenogaster vulgaris, H. hessei* and the rare *Gymnomyces xanthosporus*. Other interesting fungi recorded from Betws-y-Coed included *Inocybe auricoma, Gomphidius glutinosus,* and *Ramaria formosa*, an attractive and rare fairy club fungus. *Russula integra*, a northern species in Britain, associated with pine, was found in woodland by Llyn Ogwen, along with the scaly-capped birch associate, *Cortinarius pholideus*.

The first BMS foray to target the higher slopes of Snowdonia was during early autumn 1989. On this occasion, the group concentrated on the Snowdon range, and adopted a strategy of ascending the mountain by a different route for each day of the foray. This worked out very well, producing new records for many upland grassland species. The genera *Hygrocybe, Galerina, Entoloma* and *Omphalina* in particular were well represented. The most interesting finds though were associated with the arctic-alpine Dwarf Willow (*Salix herbacea*) community located on the summit ridge. Two truly montane species, *Galerina hypophaea* and *Omphaliaster borealis* were found and identified by Professor Roy Watling of The Royal Botanic Garden, Edinburgh. Also with the Dwarf Willow were *Naucoria bohemica* and *Laccaria bicolor*, fungi normally found with forest trees in lowland areas.

In addition to the BMS forays, the North Wales Wildlife Trust has been holding its own autumn forays for over thirty years. They were led in the 1960s and early 1970s by Dr Geoffrey Dobbs, a mycologist working at Bangor University, and subsequently by Nigel Brown, curator of Treborth Botanic Gardens. They have mostly concentrated on woodland sites, such as the oak woods around Capel Curig and the coniferous plantations of Coed Gwydir, especially at Nant Bwlch-yr-Haiarn. These local groups have tended to focus on the larger and more conspicuous fungi, but they were hampered for many years by the lack of reliable field guides. Over the past decade though, the situation has improved greatly, and fungi such as the Russulas, *Lactarii*, and boletes have begun to be much better recorded. Among the uncommon species they found were the pale bolete *Leccinum holopus*, and the large yellowish milk cap *Lactarius repraesentaneus*, both from *Sphagnum*-rich birch habitats in Coed Padarn. From Capel Curig the fleshy milk cap *Lactarius volemus*, a declining species, was occasionally found under oak and, from Nant Bwlch-yr-Haiarn, *Tricholoma psammopus* was recorded under Larch.

At the present time areas of upland Snowdonia remain unsurveyed by mycologists. While the valley woods have been frequently visited, mountain slopes other than on the Snowdon massif have had much less attention, and may well bring new records. This applies particularly to the alpine Dwarf Willow communities, which appear to be surprisingly rich in mycorrhizal species.

IMPORTANT HABITATS FOR FUNGI IN SNOWDONIA

Over the past three thousand years man and his livestock have created a complex landscape of artificial, semi-natural and natural habitats in Snowdonia, and while today's Atlantic climate still favours oak woodland on all but the highest ground, human influence has created vast tracts of acid grassland and heathland. The fragmentation of Snowdonia's woodlands may have resulted in the loss of certain woodland fungi, but fortunately many of the semi-natural habitats that replaced them now support rich mycofloras of their own.

HIGH ALTITUDE HABITATS

On mountainsides, areas of bare impoverished soil or exposed peat, especially on the upper slopes, are a habitat for the unusual lichenised *Gerronema* species. This is a basidiomycete genus, but includes species that form symbiotic partnerships with algae, to produce what appear to be typical encrusting lichens. They are atypical though in the fact that they produce umbrella-shaped fruiting bodies instead of the usual cup-shaped structures of the much more numerous 'ascomycete' lichens. Snowdonia has three species of *Gerronema, G. ericetorum, G. hudsonianum* and *G. alpinum*, that are known to be lichenised.

The miniature alpine woodlands of Dwarf Willow found on the summits of mountains such as Elidir Fach are another important habitat for mountain fungi. As previously mentioned, alpine species like *Galerina hypophaea* and *Omphaliaster borealis* can be found here, and some, such as *Naucoria bohemica*, form a mycorrhizal association with the Dwarf Willow. A recent foray on the Glyderau has also revealed the presence of the rare mountain grisette *Amanita nivalis*, associated with Dwarf Willow together with species of *Inocybe, Cortinarius* and *Russula*, which have yet to be identified.

MOUNTAIN GRASSLANDS

We have seen that although the mountain slopes of Snowdonia have mainly been denuded of their tree cover, they are still an important habitat for fungi. The species are mostly saprophytes, growing among grass or on the dung of grazing animals. The most striking of the grassland species are the brilliantly coloured wax caps (Hygrocybes). Red species are represented by Crimson Wax Cap (*Hygrocybe punicea*), Scarlet Hood (*H. coccinea*) and the finely scaly *H. helobia*; yellow species by the slippery *Hygrocybe ceracea* and *H. chlorophana*; orange

Crimson Wax Cap (*Hygrocybe punicea*) in fields near Tregarth.

species by *Hygrocybe reidii*, and green ones by the viscid Parrot Toadstool (*Hygrocybe psittacina*). These agarics are sure indicators that fertilizers have never been applied, and such sites are often rich hunting grounds for other fungi. The Crimson Wax Cap occurs only on old unimproved pastures and is usually found with an assortment of other wax caps, perhaps including the distinctive pink *Hygrocybe calyptriformis,* which tends to occur in lusher patches of pasture, close to trees or shrubs. Another striking species is the rare *Hygrocybe spadicea,* which has been recorded from Snowdon; unusually for this group, it has a dark brown cap, which contrasts vividly with its yellow stem and gills. *Entoloma* species are well represented too, and although not as colourful as wax caps, some have distinct bluish tones such as *E. chalybaeum, E. serrulata,* and the uncommon *E. bloxamii*, recorded from Snowdon. Others though, like *Entoloma conferendum*, a common species of unimproved pastures, are distinctly drab. Liberty Cap (*Psilocybe semilanceata*), the notorius hallucinogenic species, can be abundant in unimproved pastures. It is a small dingy toadstool but has a characteristic bump, or papilla, at the apex of its cap. In patches of lush, mossy grass one may also come across groups or 'troops' of slender *Mycena* species such as *M. galopus*, a species normally associated with woodlands. Fairy club fungi such as the brightly coloured *Clavulinopsis corniculata* and *C. helvola* occur, and sometimes, low-down in the turf, the brownish *Clavaria fumosa*. Less attractive are the blackish earth tongues, which can occur in profusion and often remain intact into the winter. These curious protrusions include several species such as *Geoglossum viscosum* and *G. fallax*, which can be distinguished with certainty only with a microscope. They are saprophytic species, but the Scarlet Caterpillar Fungus (*Cordyceps militaris*) with a scarlet tongue, is a parasite, preying on the pupae of moths and butterflies beneath the soil surface. It has been seen, for example, on the slopes of Nant Ffrancon and Moel Eilio. Well-drained heathy grassland may give rise to large numbers of the scaly-stalked *Cystoderma amianthinum* together with the attractive wax cap *Hygrocybe cantharellus*, a species that is widespread in Snowdonia.

WETLAND HABITATS

The mires, marshes, fens and bogs of Snowdonia often have their own distinctive fungi. *Sphagnum* bogs, for example, usually contain a number of *Galerina* species such as *G. tibiicystis* and *G. paludosa*. Other species of the habitat include the greyish, cucumber-smelling *Tephrocybe palustris,* which usually appears during the summer months, and a variety of brightly coloured wax caps, including *Hygrocybe turunda* and *H. coccineocrenata* that are more or less restricted to *Sphagnum* bogs. Areas of wet peat, such as those on the slopes of Cwm Idwal, can support a variety of *Hypholoma* species. These small yellowish or reddish brown toadstools are closely related to the familiar woodland Sulphur Tuft (*Hypholoma fasciculare*), but grow in an entirely different habitat. *Hypholoma elongatum* is the commonest, and characterised by a long stipe which roots deeply into the substrate, while *Hypholoma ericaeum* is shorter with a more reddish cap.

Galerina tibiicystis on Snowdon.

Sulphur Tuft (*Hypholoma fasciculare*) in Bryn Meurig.

Many marshland fungi are found on the dead stems or leaves of their specific hosts. The dead stems of rushes, for example, are host to myriads of tiny saprophytic fungi. Minute hairy cup fungi of the genus *Lachnum* such as *L. apalum* and *L. diminutum* are also likely to be encountered. These are beautiful fungi, but can only be fully appreciated under a low-power microscope. Another, somewhat larger cup fungus, is *Myriosclerotinia curreyana*, which produces clusters of brown cups at the base of rush tussocks. They arise from hard black structures (sclerotia) embedded in the remains of the dead stems. Another ascomycete of wetland habitats is the brightly coloured Bog Beacon (*Mitrula paludosa*), a characteristic species of *Sphagnum* bogs and submerged leaf litter.

Bog Beacon (*Mitrula paludosa*) in marshes near Tregarth.

Many cup fungi are also found on sedges, and one of the more conspicuous ones is *Myriosclerotinia sulcata*, which has cups arising from a peculiar banana-shaped sclerotium. It has been found, for example, on sedge remains at the edge of Llyn Bodgynydd. Stands of Purple Moor-grass also conceal many microfungfi on their dead stems, *Belonium hystrix* being one of the better-known ones. It is a tiny cup fungus ornamented with short dark bristles. *Molinia* stems are also likely to be dotted with numerous black specks of *Actinothyrium graminis*, an asexual microfungus. Cotton-grasses, spike-rushes and others are also rich substrates for fungi, but have not been investigated in any detail in Snowdonia.

WOODLANDS

Woodlands provide more niches for fungi than any other habitat, so it is not surprising that most mycologists give them a lot of attention. Ground floras usually include a large number of basidiomycetes, often in a complex mixture. There may be saprophytes such as *Mycena* spp. (Bonnet Caps), *Collybia* spp. (Tough Shanks) and *Clitocybe* spp. (Funnel Caps); and mycorrhizal species, such as *Lactarius* spp. (Milk Caps), *Russula* spp. (Brittle Gills) and various boletes. There are also likely to be parasites such as the notorious Honey Fungus (*Armillaria mellea*) which can kill apparently healthy trees, and the much smaller leaf-dwelling mildews and rusts.

Honey Fungus (*Armillaria mellea*) near Moel y Ci.

OAK WOODLANDS

Many of the spectacular oak woodlands of Snowdonia, such as Coed Crafnant, Coed Camlyn, Coed Cymerau, Coed y Rhygen, and especially Coedydd Aber, have outstanding fungus floras. Milk Caps, so called because they exude white or coloured latex when damaged, are usually frequent, and many are associated with oak. By far the most common species is the Oak Milk Cap (*Lactarius quietus*). *Lactarius volemus* by contrast now seems to occur only in the woodlands around Capel Curig, but was more widespread at one time. More common and certainly more conspicuous is the Fleecy Milk Cap (*L. vellereus*), with a cap sometimes achieving a diameter of 25 cm. Most species produce white latex, but the latex of *L. chrysorrheus*, which occurs occasionally in acidic oak woodlands such as Bryn Meurig, turns a beautiful sulphur yellow moments after it is exposed to the air. Another unusual milk cap of oak woodland is *Lactarius fuliginosus*. It has a very dark cap with a velvety texture and gills that turn reddish when damaged. It occurs sporadically in Coed Padarn and at one or two other sites in Snowdonia.

The Russulas are another common group of oak woodlands. They are close relatives of milk caps, but produce no latex and have rather fragile gills, hence their collective common name of brittle gills. They also come in a much greater variety of colours than the Milk Caps. One of the commonest is the the Charcoal Burner (*Russula cyanoxantha*), so called because of its variable cap colour – sometimes lilac, greenish with dark grey, ashen, or even completely green (variety *peltereaui*). Other common species are the Blackish-purple Russula (*R. atropurpurea*) with a dark red shiny cap, the Common Yellow Russula (*R. ochroleuca*) with dull yellow or olive-yellow colours, and the small, easily overlooked *R. lutea*, which is viscid with orange tones. One of the largest species, the Blackening Russula (*R. nigricans*), is drab brown at first, but turns black as it ages. *Russula emeticella*, a small, fragile species with a bright red cap, is rather common in the oak woods of Capel Curig. It is a hot-tasting species of the 'Sickener' group, closely related to the more robust pine forest species *Russula emetica*, known as the Sickener because of its emetic properties. Some of the Russulas are nondescript with respect to colour, but easily identified by their aromas: marzipan-like in *Russula laurocerasi,* and a less pleasant oily-fruity smell in the Foetid Russula (*R. foetens*).

Oak woodlands are also a habitat for the enigmatic Amanitas. Several of them are very poisonous, and it is mainly because of them that toadstools have such a bad name. The fruiting bodies have a more complex development than in the brittle gills or milk caps, and start life shrouded in protective veils. To reach maturity they have to break free of these veils, and this usually leaves a residual ring or annulus on the stipe (or stem) and a cup shaped structure at the base of the stipe known as a volva. The most deadly of the amanitas is the Death Cap (*Amanita phalloides*), sinister olive green in colour and probably responsible for more human deaths than any other fungus. Perhaps fortunately, it is rare in Snowdonia, with records only from Halfway Bridge (Coed Cochwillan) between Bangor and Bethesda. It is much more common, though, on the alkaline soils of Ynys Môn (Anglesey) and in the Afon Menai (Menai Strait) area. The closely related False Death Cap (*Amanita citrina*) is much commoner in Snowdonia (around Tan y Bwlch for example), but fortunately harmless. It usually has a pale yellow cap, although a completely white variety (var. *alba*) occurs, which could be confused with the deadly Destroying Angel (*Amanita virosa*). The False Death Cap is distinct, though, in its rounded volva and distinctive smell of raw potatoes, and to date the Destroying Angel has not been found in Snowdonia.

Another highly poisonous and striking species is the Panther Cap (*Amanita pantherina*) with a dark brown cap, speckled with pure white scales. It is rare in Snowdonia, having been recorded only from the woods around Bethesda. Much more common is the edible Blusher (*Amanita rubescens*) distinguished by a more pinkish hue to the cap and stem, and a pink bruising reaction. *Amanita porphyria* is a greyish species, which has occasionally been found in mixed woodland with oak, such as Coed Bryn-Engan, near Capel Curig, and recently at Betws-y-Coed growing under conifers. Because of possible confusion between the edible and poisonous species of *Amanita*, the whole group should be avoided for culinary use.

The boletes are a large group of fungi distinguished by the fact that they have pores instead of gills. A common species of Snowdonia's acid oak woodlands is *Boletus erythropus,* which has distinctive orange-red pores which quickly turn blackish on bruising. A beautiful but scarcer species is *B. calopus* which has a bright red stipe overlain by pale netting. Both are recorded in the oak woods at Capel Curig and may be accompanied by the occasional specimen of the drab-coloured *Porphyrellus pseudoscaber* – a species also known from other places in Snowdonia. The stocky *Boletus pulverulentus*, which blackens on handling, was recorded from the woods near Llyn Ogwen on the 1950 BMS foray, and has been seen more recently near Maentwrog and at Caerhun in Dyffryn Conwy. One of the more bizarre boletes is the rare *Xerocomus parasiticus*, which appears to be a parasite of another fungus called the Common Earth Ball (*Scleroderma citrinum*). It is not quite clear whether this is an example of true parasitism, but *X. parasiticus* is never found without this particular host. It is not uncommon in the oakwoods of Snowdonia and, at one site in Nant Ffrancon was found in association with several earth balls in the same area. The host fungus, a yellowish, tough-skinned gasteroid (stomach-fungus), is almost ubiquitous on

Xerocomus parasiticus on Common Earth Ball (*Scleroderma citrinum*) in Nant Ffrancon.

Beefsteak Fungus (*Fistulina hepatica*) in Bryn Meurig.

Many-zoned Polypore (*Coriolus versicolor*) in Coedydd Aber.

well-drained, mossy turf in oakwoods. Also in this habitat, one may find the club-shaped *Cordyceps ophioglossoides* – initially with a greenish hue but blackening with age. Careful excavation reveals that it too is parasitic – in this case on the Hart's Truffle (*Elaphomyces muricatus*).

The oak trees themselves are also host many fungi. Brackets such as the strange flesh-like Beefsteak Fungus (*Fistulina hepatica*) can seen in Bryn Meurig, for example, and the brightly coloured Sulphur Polypore (*Laetiporus sulphureus*) at Coed Cochwillan. These two polypores generally grow as weak parasites, but the common Many-zoned Polypore (*Coriolus versicolor*) and the Oak Maze Gill (*Daedalea quercina*) are entirely saprophytic, growing on stumps and fallen trunks. The dead branches are also worth investigation and may carry the liquorice-like fruiting bodies of the Black Bulgar (*Bulgaria inquinans*). Brick Caps (*Hypholoma sublateritium*), a larger version of the Sulphur Tuft and named after its colour, occurs occasionally on oak stumps. Another large fungus often associated with oak, is *Megacollybia platyphylla*, a toadstool characterised by tough rhizoids which penetrate the rotten wood of dead stumps.

BEECH WOODLANDS

Though not native to North Wales, Beech has been widely planted in the valleys of Snowdonia and there are many species of fungi associated with it. Common among the mycorrhizal fungi is the Geranium-scented Russula (*Russula fellea*), which occurs, for example, in the beechwoods near Ganllwyd. The bright red Beechwood Sickener (*Russula mairei*) can also be plentiful, along with the Common Yellow Russula (*Russula ochroleuca*). Much scarcer is *Russula lepida*, looking similar to the Beechwood Sickener, but less fragile, and with a distinct smell of cedarwood pencils. It has been recorded from Aberglaslyn and in Beech woodland near Bethesda. The Slimy Milk Cap (*Lactarius blennius*) is almost ubiquitous under Beech trees, but another species, the pale *L. pallidus* has its only Snowdonia record from Coed Dolgarrog, perhaps because of a requirement for nutrient-rich soils. A viscid agaric common under Beech is the violet-stemmed *Cortinarius stillatitius*, formerly known as either *C. elatior* or *C. pseudosalor*. Boletes such as *Boletus erythropus* and *B. calopus* can occur on acid soil under Beech trees as well as oaks, and the common Red-cracking Bolete (*Xerocomus chrysenteron*) is frequently found in such places. One of the most distinctive saprophytic toadstools of Beech is the glutinous Porcelain Fungus (*Oudemansiella mucida*). Its almost ivory white fruiting bodies can produce spectacular displays, often high up on the trunks and branches of dead standing trees. Fallen branches are often colonised by the black and somewhat sinister-looking ascomycete Dead Man's Fingers (*Xylaria polymorpha*). It may be accompanied by the purple gelatinous *Ascocoryne sarcoides,* another spore shooter, which starts off looking like inverted cones, but eventually produces cup-shaped structures. On the same substrate but much less common is *Neobulgaria pura*, which forms clusters of pinkish barrel-shaped fruiting bodies when young. Its only known localities in Snowdonia are at Betws-y-Coed and in Coed Dolgarrog.

Beech litter is a habitat for many saprophytic fungi. One of the most beautiful is the Amethyst Deceiver (*Laccaria amethystina*), so-named because its colour fades to a pale lilac on drying. Half concealed by the litter one may also notice the pale clumps of the Hedgehog Fungus (*Hydnum repandum*), a common edible species with spines rather than gills. It is distinct from the darker coloured spined fungi – in the genera *Phellodon* and *Hydnellum* – which are much rarer. In Beech woods, one may also see a slender relative of the Chanterelle called *Cantharellus tubiformis* which, like the Chanterelle, is good to eat.

Cantharellus tubiformis near Rhaeadr Ewynnol (Swallow Falls).

A good place for it is the woodland around the Miner's Bridge near Betws-y-Coed. A much smaller and dingier and consequently easily overlooked member of the family is *Pseudocraterellus sinuosus*, which forms tight clusters under Beech and oak. It could not be described as common in Snowdonia, but does occur quite widely – in the woodlands around Capel Curig for example, and in Coed Padarn. A species which is always a pleasure to find is the curious Jelly Babies (*Leotia lubrica*), a stalked ascomycete with rubbery fruiting bodies. It often occurs gregariously on soil under Beech trees. One of the more common bracket fungi of Beech is the parasitic *Ganoderma applanatum*, which attacks the tree and causes white rot. It is a perennial species and over the years forms great tiers of shelf-like fruiting bodies.

CALCAREOUS WOODLANDS

Lime-rich soils occur only very locally in Snowdona, and the fungi associated with them are consequently scarce. One of the best examples of calcareous woodland is Coed Dolgarrog in Dyffryn Conwy, and several lime-loving fungi occur there. They are often associated with deciduous trees, particularly Beech, and may include several members of the parasol group (*Lepiota* and allied genera). The rarest is *Chamaemyces fraccidus*, which was recorded on the 1988 BMS foray, but has not been seen since. It is a fungus characterised by a brown scaly stem which exudes yellowish droplets. Also distinctive is *Lepiota ignivolvata* which has been recorded here on three occasions but at no other site in Snowdonia. It is a medium-sized whitish species with a distinctive orange girdle on the stem. The small powdery toadstool *Cystolepiota bucknalli* with its unique combination of lilac colours and a strong gaseous smell also occurs in this locality.

BIRCH WOODLANDS

Birch woodlands generally have a rich mycoflora. The most distinctive of the larger mycorrhizal species is the Fly Agaric (*Amanita muscaria*), the legendary red spotted toadstool of fairytale fame. In some years it can occur in profusion, with some specimens achieving dinner-plate dimensions. It was used by the Lapps as a hallucinogen and appears to have played an important part in tribal religion and culture. It is poisonous though, and has caused deaths. Its common

Fly Agaric (*Amanita muscaria*) in Coedydd Aber.

name comes from a one-time use as a fly-deterrent in which the flesh was broken up and added to saucers of milk. Another common (but edible) species, the Tawny Grisette (*Amanita fulva*) is also a birch associate and often favours sites with a ground-layer of Bilberry. The greyish Grisette (*Amanita vaginata*) is less common but still widespread in Snowdonia. *Amanita crocea*, a beautiful orange-coloured species, is one of the scarcer Grisettes, but does occur quite frequently in parts of Coed Padarn.

Milk caps can be common in birch woods, especially in the damper areas. Among the species are *Lactarius tabidus*, which produces white milk which quickly turns yellow on contact with a white handkerchief; the robust dark green Ugly Milk Cap (*L. turpis*); the pallid but strongly aromatic Coconut-scented Milk Cap (*L. glyciosmus*); and the Grey Milk Cap (*L. vietus*) which has similar colours to *L. glyciosmus* but little smell and often grows among *Sphagnum* moss. Some species, such as the common Woolly Milk Cap (*L. torminosus*), have a distinctive shaggy fringe to the cap margin. This is also a feature of a much rarer species of boggy-ground *L. repraesentaneus,* which has been found in the woodlands adjacent to Llyn Padarn, but in Snowdonia is near the southern limit of its British range. *Lactarius uvidus*, another northern milk cap, was recorded recently from Llyn Bodgynydd on damp ground under birch and sallow. As with *L. repraesentaneus*, the gills react to damage by turning purple.

There are a number of different *Cortinarius* species that grow on acid ground under birch including *Cortinarius armillatus* with a red-girdled stipe, and the very viscid *C. delibutus*. Both have been found among Bilberry in Coed Padarn. The scaly, aromatic *C. paleaceus* also occurs here, although it can grow under

Cortinarius paleaceus in Tregarth woods.

pine in other places. Very common among the Russulas found under birches is the Common Yellow Russula (*R. ochroleuca*), but on damper ground it is sometimes replaced by the brighter and more viscid Yellow Swamp Russula (*R. claroflava*). In such conditions this species may be accompanied by the reddish *R. nitida*, which is at home among *Sphagnum* moss. Also likely in this habitat are the extremely fragile, pallid-pink caps of *R. betularum*.

Yellow Swamp Russula (*Russula claroflava*) in Tregarth woods.

The boletes of birch woodlands are mainly represented by species of the genus *Leccinum*. Early in the season, especially on the better-drained soils, the Brown Birch Bolete (*L. scabrum*) can occur in large numbers, but it is usually replaced by the blue-green staining *L. variicolor* or the pale, slender *L. holopus* on boggy ground. On heathy ground, the Orange Birch Bolete (*L. versipelle*) can provide a contrasting spash of colour. Among the rarer boletes, the pale beige *Gyroporus cyanescens* has recently turned up in the woods at Capel Curig – only the second record for Snowdonia, following one from Fairy Glen in 1924.

Dead standing birch trees and their fallen trunks often display outgrowths of the large, rubbery brackets of the Razor-strop Fungus (*Piptoporus betulinus*), the common name referring to the fact that the dried brackets were once used for sharpening razors. Unlike many bracket fungi, it is an annual species and can become host to other saprophytic fungi as it ages. The tiny orange ascomycete *Nectria peziza* is one of several possible colonizers.

Other bracket fungi include the Hairy Stereum (*Stereum hirsutum*), the Many-zoned Polypore Bracket (*Trametes versicolor*) and the Tinder Fungus (*Fomes fomentarius*), but although the last species is common in Scotland, so far there is only a single record for Snowdonia (at Pensychnant). Another rare bracket fungus associated with birch in Snowdonia is *Lenzites betulina*. It resembles the much more common Blushing Bracket (*Daedaleopsis confragosa*), which may also occur on birch, but differs in having gill-like structures rather than the more usual pores. The only recent record is from Coed y Rhygen, where it was found in 1979. *Coltricia perennis* is among a group of often perennial fungi with pores like those of many bracket fungi, but with a cap held clear of the substrate by a central stalk. It differs from most of its relatives in the fact that it grows on soil rather than wood, and it has been found recently under birch in Coed Padarn, although it is also known from other habitats.

PINEWOODS

Although the pinewoods of Snowdonia are not truly native, they still support an impressive array of mycorrhizal fungi, with both Scots Pine (*Pinus sylvestris*) and Corsican Pine (*Pinus nigra* var. *maritima*) contributing. *Suillus bovinus* is a common bolete that often occurs in dense groups under Scots Pine, and growing with it in a few places (e.g. Coed Gwydir), one may find the attractive Rose Spike Cap (*Gomphidius roseus*), which appears to depend on the presence of *S. bovinus*.

Suillus bovinus in Nant Ffrancon.

Other common boletes include the highly viscid Slippery Jack (*Suillus luteus*) and *S. variegatus*. The Russulas associated with pine are often brightly coloured in purple or red: *R. sanguinea, R. sardonia* and *R. emetica* are typical examples. The distinctive (and edible) Saffron Milk Cap (*Lactarius deliciosus*) may be present, but in more acidic woods it is often replaced by *Lactarius quieticolor* – a similar species but with more brownish tones. On more open and heathy pinewoods, the Rufous Milk Cap (*Lactarius rufus*) becomes the characteristic species. The strange *Cordyceps capitata* may also be found among mosses or needle litter in these more open pinewoods, and like its close relative *C. ophioglossoides*, it also parasitizes Hart's Truffles, usually *Elaphomyces granulatus*.

Cordyceps capitata in Nant Ffrancon.

Saprophytes of pinewoods often include the magnificent gold-tawny toadstools of *Gymnopilus junonius*, and the bright yellow Stag's Horn Fungus (*Calocera viscosa*). Slightly less common is the curious Jelly Tongue (*Pseudohydnum gelatinosum*), which occurs on fallen trunks of pine and other conifers. One of the most spectacular species, though, is the Cauliflower Fungus (*Sparassis crispa*), which occurs at the base of living pine trees. This attractive and aromatic fungus does resemble a cauliflower, at least at a distance, and, although it may not taste exactly like one, it is good to eat. The extensive pine plantations around Nant Bwlch-yr-Haiarn are good places to see these fungi, especially at the forest edge.

Fungi are among the few organisms that can tolerate the conditions inside dense conifer plantations. The Common Yellow Russula for example, actually thrives in them, as do some *Inocybe* species (*I. calamistrata*, for example) and saprophytes such as Crested Coral Fungus (*Clavulina cristata*). One may also see clusters of the dark brown cup fungus *Peziza badia* in very dark situations.

Towards the margins of plantations, especially where they merge into birch woods, mycorrhizal fungi often become more frequent. The attractive orange-gilled *Cortinarius cinnamomeus*, for example, is a characteristic species of mixed spruce and birch. The Brown Roll-rim (*Paxillus involutus*) is common in coniferous rides, often but not always found where there are birches. Although the species has gills, microscopic characters show that it is actually related to the spongey-capped boletes. It is also poisonous, unlike the highly-prized Cep (*Boletus edulis*), which often turns up at the edge of conifer plantations. At Nant Bwlch-yr-Haiarn Ceps can occur in large numbers, usually in the company of another edible species, the Bay Bolete (*Xerocomus badius*).

Peziza badia near Rhaeadr Ewynnol (Swallow Falls).

Although in general plantations are not good for rarities, the bolete *Pulveroboletus lignicola*, previously known in Snowdonia only from Llanberis, has recently been recorded in a coniferous plantation near Betws-y-Coed. It is unusual among boletes in the fact that it grows on rotting wood rather than soil.

ALDER AND SALLOW CARR

Although Alder woods have a distinctive mycoflora, compared to other trees they have rather few associated species, and many of them are small and dingy. Characteristic of the habitat are small toadstools of the genus *Naucoria* such as

N. escharoides and *N. subconspersa*. There is also a small milk cap, *Lactarius obscuratus*, which is probably not uncommon in Snowdonia, though further survey would be needed to confirm this. Another wet-woodland milk cap, so far found only in Coed Padarn, is the pale, purple-staining *L. aspideus*. The *Cortinarius* species found with Alder include *C. alnea* and *C. helvelloides*, and there are records of both species from Coed Padarn, and of *C. helvelloides* from Coed Ganllwyd, near Dolgellau. The red-capped *Russula pumila* is one of the few Russulas to be found associated with Alder and, although we have only one locality to date – Coed Padarn – it may well be widespread.

There are some fine stands of Alder in Coedydd Aber, and it is a good place to see the attractive bracket fungus *Inonotus radiatus* growing on dead, but still standing trunks. It is recognisable by the silvery sheen of its pore surface. *Mycena haematopus*, a bonnet cap which bleeds when damaged, is also commonly found here, again on dead wood. In the springtime, the little bright yellow ascomycete fungus known as the Bog Beacon (*Mitrula paludosa*) can be found on the wet litter and among *Sphagnum* moss.

One of the most distinctive fungi of sallow carr is the Blushing Bracket (*Daedaleopsis confragosa*), a species that usually grows on dead sallow branches, but has also been found on birch in Snowdonia. Its common name refers to the fact that the undersurface readily bruises pink on handling. A less prominent bracket found on sallows is the dainty *Polyporus varius* var. *nummuliformme*, which has a beige-coloured cap and contrasting blackened stipe. By lakeshore sallows on Llyn Padarn is the slimy but beautiful *Cortinarius trivialis*, with a distinctively girdled stipe. Equally striking is *Cortinarius uliginosus* with a bright orange-red cap, found recently in Coedydd Aber.

ASHWOODS

Ashwoods stand apart from other woodlands in that they are mainly the home of saprophytic rather than mycorrhizal fungi. One of the most distinctive is King Alfred's Burnt Cakes (*Daldinia concentrica*), an ascomycete that forms round and bone-hard, fruit bodies on dead trunks and branches. These are cocoa-brown when young and fertile, but turn jet-black as they age. The gilled bracket fungus known as the Soft Slipper Toadstool (*Crepidotus mollis*) may also be present. Other smaller species of the genus *Crepidotus* (*C. luteolus*, for example) occur on fallen twigs, but they are identifiable to species only with a microscope. Saprophytic toadstools of the genus *Pluteus* also occur on Ash wood and include unusual species, such as the greenish *P. xanthophaeus*, recently recorded from Coedydd Aber. A parasitic species, *Pholiota squarrosa*, also occurs on the Ash trees at Coedydd Aber, appearing as large clumps of scaly toadstools clustered around tree bases. Surprisingly, it does not kill the tree, but appears year after year on the same tree – often persisting into the late autumn. On fallen branches or sometimes on branches that are still attached to the tree, growths of the unusual gelatinous Yellow Brain Fungus (*Tremella mesenterica*) may be encountered.

Yellow Brain Fungus (*Tremella mesenterica*) in Bryn Meurig.

CONSERVATION OF FUNGI IN SNOWDONIA

As with flowering plants, fungi thrive best in undisturbed ecosystems and a rich mycoflora is often accompanied by a rich diversity of other organisms. Like some flowering plants, fungi are intolerant of fertilisers, and pastures that have been improved for agriculture are generally much poorer in species than unimproved ones. Conversely, some of our unimproved pastures, such as the patchwork of sheep pastures close to Brynrefail, near Llanberis, are so rich in *Hygrocybe* species that they are sometimes referred to as wax cap meadows. Application of fertilisers to these meadows would change the species composition of both flowering plants and fungi in favour of more nitrogen-loving species. Only a few common fungi such as *Panaeolus rickenii* and *Conocybe tenera*, and possibly a few that grow on dung like the Dung Roundhead (*Stropharia semiglobata*), can then thrive. Recovery after a single fertiliser application can take many years.

Fortunately, there are still large tracts of upland grassland in Snowdonia where a combination of traditional farming practice, difficult terrain and climate have allowed fungi to flourish, and these now form an important part of our national heritage. Similarly, the undisturbed natural woodlands of Snowdonia, with native trees such as oak and birch, hold a huge diversity of fungi. Plantations, by comparison, are less productive and usually include only the more common species. Also, there is concern that many of the woodlands in

Snowdonia are losing species because of colonisation by Rhododendron (*Rhododendron ponticum*). This aggressive species can eliminate a natural ground flora and convert even a rich habitat such as oak woodland into a fungus desert. Mycorrhizal fungi are particularly vulnerable.

The importance of Snowdonia as a refuge for rare fungi is clear from the fact that its flora includes twelve species on the Red List of British fungi, which forms part of the World Conservation Union list of endangered and vulnerable species. They are *Amanita nivalis, Entoloma bloxamii, Hydnellum ferrugineum, H. scrobiculatum, Hygrocybe calyptriformis, H. spadicea, Phellodon confluens, P. melaleucus, P. niger, Pseudocraterellus sinuosus, Pulveroboletus lignicola,* and *Russula pumila*.

ADDITIONAL READING

Lacey, W. S. (ed.) 1970. *Welsh Wildlife in Trust*. North Wales Naturalists' Trust, Bangor.

Lacey, W. S. & Morgan, M. J. (eds.) 1989. *The Nature of North Wales*. Barracuda Books Ltd, Buckingham, England

Lockley, R. M. 1970. *The Naturalist in Wales*. David & Charles, Newton Abbot.

Ramsbottom, J. 1977. *Mushrooms and Toadstools*. The New Naturalist. Collins, London

LICHENS

ALLAN PENTECOST

INTRODUCTION

Lichens are made up of two components, a fungus and an alga, living in association to produce a plant that bears little resemblance to either partner. Both fungus and alga are thought to benefit from the association, the alga by releasing nutrients produced by photosynthesis, and the fungus by protecting the alga physically against extremes of temperature and other environmental factors, and possibly releasing mineral salts from the substrate. The result is that both alga and fungus are able to occupy habitats that would otherwise be too hostile for them in their free-living state, and possibly even too hostile for any other organism. They are found, for example, in the extreme cold of the Antarctic and the extreme drought of the Namibian desert, and may well have been the first colonisers when life first moved onto the land some 400 million years ago.

Lichenologists classify lichens in to two groups, the macrolichens and microlichens. Macrolichens are conspicuous, forming large leafy rosettes, beard-like tassels, or large crusts on trees and rocks, and were the first lichens to be noticed by the early botanists. The Crottle (*Ochrolechia tartarea*) was scraped from rocks and trees in Wales in former times to make a red dye: a process in which the powdered lichen was steeped in urine, made into dye balls with lime, and then added to boiling water containing the cloth. Only a few lichen species have been collected to the point that their survival was in question. The microlichens, as their name indicates, are less conspicuous, some being visible only with the aid of a hand lens, and identifiable only with a microscope. It has meant that, although they include far more species than the macrolichens, they have been comparatively neglected. Even now, their distribution is incompletely known, and additions to the British flora will certainly continue well into the third millennium.

Both the fungal and algal cells of lichens are contained within a plant body called a *thallus*. This bears reproductive organs, together with structures that help it cling to the substrate. The thallus of a macrolichen is either foliose (leafy) and attached by root-like structures called *rhizinae*, or fruticose (beard-like) and attached by pads of fungal tissue. If grown in isolation, most microlichens would form circular crusts, but they usually have to compete with other lichens, and this often creates irregular and colourful mosaics. The thallus is often so closely attached to the substrate that it can't be separated without damage.

The fungus component of most lichens is capable of reproducing sexually. Spores are produced in special sac-like bodies called *asci*. These are very small but grouped in structures called *apothecia,* which are quite noticeable. Most are disc-shaped but they can also be flask-shaped and slit-shaped, and are important in lichen classification. The apothecia are normally conspicuous and abundant and when a lichen produces apothecia it is said to be 'in fruit'. At this stage the dust-like spores are often forcibly ejected, and can be carried by wind for great distances. If one settles in a suitable place, it will germinate to form hyphae, but must still meet a suitable algal partner if it is to develop into a new lichen thallus. However, when the thallus is fully formed, the algal component is not usually obvious. It can provide a green tinge to some species, but tends to be masked by fungal hyphae and the Latin names given to lichens are based only on the fungal component. Several groups of fungi are involved, but the ascomycetes are by far the most important. Basidiomycete ('toadstool') lichens, such as *Gerronema hudsonianum,* do exist but they are uncommon in Snowdonia.

In addition to sexual reproduction, many lichens can undergo asexual reproduction through propagules containing both algal cells and fungal hyphae. These may simply be broken fragments of thallus, or they may be formed in special structures known as *soralia*, which produce small powdery propagules (*soredia*). In other cases, they produce larger propagules with a firm outer cortex, known as *isidia*.

THE HISTORY OF LICHEN EXPLORATION IN SNOWDONIA

Lichens have been studied in Snowdonia over a long period. The earliest published record appears in John Ray's *Synopsis methodica stirpium Britannicarum* (1696), and is of *Cladonia furcata*, a macrolichen collected by Edward Lhuyd on Carnedd Llewelyn. Lluyd made several visits to the area, noting some lichens that were later listed by Johann Jacob Dillenius in his book *Historia Muscorum* (1742) and by Gibson in *Camden's Britannia* (1789). Samuel Brewer, who spent some time in the area in 1727, also collected lichens, sometimes with Dillenius, and about 20 species, including *Sticta limbata* and *Umbilicaria polyrrhiza* are listed for the area in the *Historia Muscorum*.

In the early 19th century, four lichenologists were active in Snowdonia: John Wynne Griffiths, Dawson Turner, William Wilson and Hugh Davies. Dawson Turner visited Snowdon in 1802, perhaps accompanied by Sir James Edward Smith. The species found on the trip are listed in *English Botany*, a 36-volume work published by Sir James between 1790 and 1814. William Wilson of Warrington, although mainly interested in mosses and liverworts, collected lichens on at least seven occasions in North Wales between 1824 and 1838. He did not publish details of his finds, but his collections are in the British Museum (Natural History). John Wynne Griffiths also collected during the period and contributed to the *Botanist's Guide through England and Wales* published by Dawson Turner and Lewis Weston Dillwyn in 1805. However, the most important contribution was that of the Rev. Hugh Davies. Davies lived on Anglesey but often collected in

Lichen encrusted rocks on Glyder Fach.

Snowdonia. His records can be found in his 1813 publication *Welsh Botanology*. A few years later, the Rev. Thomas Salwey of Oswestry made some important finds in the southern part of Snowdonia near Abermaw (Barmouth) including *Physcia clementei* and *Thamnolia vermicularis*.

    As a result of these early studies, it became clear that North Wales was a rich hunting ground for lichens. Discoveries both on the continent and in Britain prompted many visits to the area by Englishmen, no doubt encouraged by the completion of Thomas Telford's road to Holyhead. Pre-eminent among them was the Rev. William Allport Leighton of Shrewsbury. Unlike his predecessors, he undertook a systematic study of the Snowdonia lichens, and discovered a large number of microlichens, some new to science. Leighton made at least 14 visits between 1856 and 1879 and published most of his findings in the *Lichen-flora of Great Britain* (1879). However, the most comprehensive list of lichen records compiled for Snowdonia in the 19th century appears in the *Flora of Anglesey and Carnarvonshire,* published by the Bangor pharmacist J. E. Griffith in 1895, and includes many of Griffith's own records. Thereafter, lichen recording in Snowdonia went into a decline until the 1920s when Walter Watson and Daniell A. Jones, among others, added further records. Watson was one of the few practitioners of

British lichenology between 1920 and 1950. Although he resided in Somerset, he travelled widely in England and Wales, and some of the arctic-alpine lichens that he recorded are mentioned by Watson in his review 'The bryophytes and lichens of arctic-alpine vegetation', published in the *Journal of Ecology* for 1925. Watson listed moss and lichen associations and compared the flora of the British mountains with that of the Alps. He noted that the higher diversity of the Alps is due in part to the greater development of calcareous rocks there. In general, however, this period was a low point for British lichenology, and it was not until after World War II that interest was revived. Over the past 50 years much has been added to our knowledge of the Snowdonia flora. During the 1960s, Kenneth Kershaw made a study of the ecology of woodland lichens in Wales, and showed how the corticolous lichen species (those that grow on tree bark) changed as the tree aged.

In 1980 the present author published the results of an ecological study of montane lichens in the *Journal of Ecology,* and discovered that rock-type was important for lichen colonisation. While this was already known for the softer rocks, it was not recognised in the slower-weathering volcanic rocks such as those of Snowdonia. Stephen Clayden, a Canadian lichenologist, continued with these investigations in the 1980s. He made growth measurements on lichens that grew on slate, and investigated the dispersal and establishment of lichen propagules in mountain regions. At the same time, the botanist Peter Benoit of Abermaw (Barmouth) became interested in the group, and made a number of interesting finds including *Parmelia quercina.* The 1980s also saw the publication of a lichen flora of Gwynedd in the *The Lichenologist* by the present author. It includes new records and information on the distribution of species and how they are influenced by climate, geology and land-use.

Research continues on the ecology of the saxicolous (rock-inhabiting) lichens. In 1977, for example, permanent quadrats were set up in the Llanberis Pass to study lichen growth rates and species succession. The first phase of the study will be completed in 2002 after 25 years of observation.

IMPORTANT HABITATS FOR LICHENS IN SNOWDONIA

Almost 900 lichen species have been recorded from the county of Gwynedd, of which about 500 occur in central Snowdonia. Their various habitats are discussed in the following sections.

WOODLANDS

Although the forest cover of central Snowdonia is low, tree bark in the remaining woodlands provides a habitat for numerous lichen species, many of which grow only on bark. The well-established Sessile Oak (*Quercus petraea*) and Pedunculate Oak (*Q. robur*) woodlands provide the richest sites. Ten of the best sites in Gwynedd are in deep valley bottoms. In the north, the main sites are Coedydd Aber, Coed Cae-Awr, Dolfriog Woods, Fairy Glen Woods, Nantgwynant (Snowdon) and Capel Curig woods. In the south the woods are larger and generally richer in species, and include Coed Benglog, Coed Crafnant, Coed Ganllwyd and Nantgwynant (Cadair Idris). Most of them contain more than 100 lichen species, many regarded as characteristic of 'old forest'. The Revised Index of Ecological Continuity (RIEC), a measure of woodland lichen richness and diversity based on the number of old forest species, scores highly for all of these sites. This system was devised by the famous botanist Francis Rose, who studied many of these woodlands in the 1970s. The woods are noted for their macrolichen flora which may include *Cetrelia olivetorum*, *Lobaria amplissima*, *L. pulmonaria*, *L. scrobicularia*, *L. virens*, *Parmelia laevigata* and *P. taylorensis*.

Lobaria virens in open woodland near Ganllwyd.

Lobaria pulmonaria, in particular, has been well studied in Britain as a conspicuous old forest lichen which has disappeared from many parts of the country, particularly from eastern areas. The woods in North Wales remain a stronghold, although there has been a decline even here. Over the last 100 years, the species has disappeared from at least four woods. It is most often seen on the bark of old Ash or oak trees, but can be sparsely distributed even at the best sites. Known as the Tree Lungwort, it was used in the Middle Ages to treat lung diseases because of its resemblance to lung tissue. All the *Lobaria* species, in common with many other lichens, are sensitive to air pollution but, in Snowdonia, the recent decline in *L. pulmonaria* is more likely to be the result of forest disturbance – either clear-felling or the replacement of broad-leaved trees with conifers. Tree species with smooth bark such as Sycamore have their own specialised lichens. *Pyrenula macrospora*, *Graphis scripta* and *G. elegans* are examples; the last two named for their resemblance to written characters.

Noteworthy woods include the Fairy Glen at Betws-y-Coed and the woods below Cwm Bychan near Llanbedr. The former, occupying a steep sheltered valley, is reminiscent of some of the best sessile oakwoods in Scotland and the west of Ireland, with several strongly oceanic macrolichens such as *Mennegazzia terebrata* and *Sticta canariensis*. Coed Crafnant and Coed Gerddibluog in Cwm

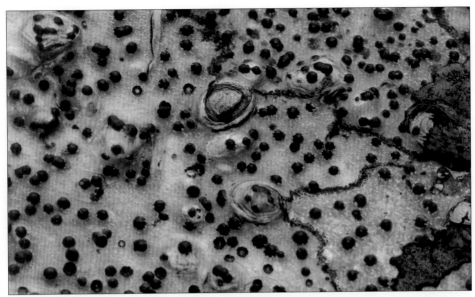

Pyrenula macrospora on Sycamore in Coedydd Aber.

Bychan have been known for their oceanic mosses and lichens since the 19th century. They have a rich and diverse lichen flora with over 150 corticolous (tree bark) species recorded in an area of less than 3 km^2. Among them are *Leptogium cyanescens, Pannaria mediterranea, Peltigera collina* and *Porina hibernica*: species that are rare or absent from other Snowdonia woodlands.

Other predominantly western macrolichens found in the woodlands of Snowdonia include *Sticta canariensis, S. fuliginosa, S. limbata, S. sylvatica,*

Sticta canariensis on stream side rocks near the Afon Mawddach.

Degelia plumbea with *Normandina pulchella* on Ash tree in Coedydd Aber.

Ochrolechia tartarea on rocks in Nant Ffrancon.

Parmelia crinita, Degelia plumbea and *Nephroma laevigatum*, all found in loose association with species of *Lobaria* on old mossy trees, forming what is known as the Lobarion community. In the more upland woodlands, the Lobarion species are replaced by others with more alpine character such as *Mycoblastus sanguinarius* and *Ochrolechia tartarea* and, particularly on conifers, *Bryoria fuscescens*. The 'artificial' conifer forests of Betws-y-Coed provide a habitat for species such as *Hypocenomyce caradocensis, Lecanora piniperda* and *Parmeliopsis hyperopta*, which are generally associated with the natural boreal forests of Scotland and northern Europe. The upland woodlands of Snowdonia, whether deciduous or evergreen, also support an abundance of beard lichens, including *Usnea ceratina, U. filipendula* and *U. florida*, and are reminiscent of Wistman's Wood on Dartmoor, which is noted for its pendulous bryophytes and lichens. Woodlands above 400 m in Snowdonia are rare, but the few that do occur (Coed Gordderu and Cae Du for example) are not outstanding for lichens, usually being dominated by common species such as *Hypogymnia physodes* and *Platismatia glauca*. It may be that they are simply too small to provide the shelter and high humidity that the more exotic lichens require.

Usnea florida on conifer tree near Rhaeadr Ewynnol (Swallow Falls).

ROCK HABITATS

The rugged heartland of Snowdonia contains a wide range of rock types supporting a large and varied lichen flora. The species may be divided according to their moisture requirements into those of dry habitats, which receive moisture only from rain, and those of semi-aquatic habitats of seepages, streams and lake margins. The dry rock species vary according the nature of the rock and fall into five groups.

1. Acid rocks

Included here are the rhyolite lavas, granites, quartzites and base-impoverished grits. They all weather slowly and form most of the high ground in Snowdonia. The flora of exposed acid rocks is dominated by crustose species. The dull purplish brown crusts of *Fuscidea lygaea* are abundant and often dominant, especially on block scree, and in combination with grey and green species such as *Fuscidea cyathoides, Porpidia tuberculosa* and *Rhizocarpon geographicum* form colourful mosaics. On rock eminences, particularly isolated boulders, the flora is usually enriched with species that are favoured by bird droppings. There is no doubt that bird manure provides nutrients to lichens and its effects are probably more important than was once thought, with even occasional manuring affecting the flora. In a study on rhyolite lava boulders in Nant Ffrancon, the yellow crustose lichen *Candelariella coralliza* was found to be completely dependent on such enrichment, and a further twelve species were recorded only from bird-perches, including the crustose species *Acarospora fuscata, Aspicilia caesiocinerea, Lecanora intricata, Schaereria fuscocinerea* (formerly *S. tenebrosa*), and the foliose species *Umbilicaria polyphylla* and *Xanthoria candelaria*. Among those that benefit from bird droppings without actually relying on them are the foliose species *Hypogymnia physodes, Lasallia pustulata* (Rock Tripe), *Parmelia omphalodes, P. saxatilis* and *Xanthoria parietina. Fuscidia lygaea,* by contrast, is a species that cannot tolerate bird droppings, but which is none the less almost ubiquitous on the acid rocks of Snowdonia and in western Britain generally. Strangely though, it is much less common in the Alps. While this species may be tolerant of a wide range of slopes and aspects, several lichens such as *Diploschistes scruposus, Fuscidea kochiana, F. recensa, Ophioparma ventosa* (formerly *Haematomma ventosum*) and *Porpidia tuberculosa* have marked preferences for steep sheltered surfaces. The last especially is one of the commonest lichens of acid rocks and noteworthy for reproducing asexually. Instead of apothecia, it produces small grey granules (soredia) containing both algal and fungal cells, and then relies on wind and rain for their dispersal. Other lichens show a preference for sunny slopes. This is particularly true of the yellow *Rhizocarpon* species (the 'map' lichens) whose bright almost fluorescent colours are often conspicuous on acid rocks and slates. The yellow pigment is produced by the fungus and forms only in the upper part of the crust – perhaps as a protection against periods of intense radiation. The colour contrasts strongly with the black of the apothecia and thallus margin.

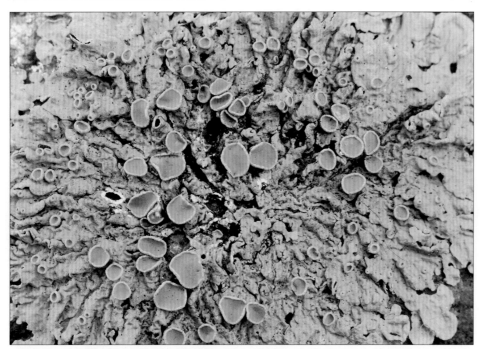

Xanthoria parietina on rocks near Moel y Ci.

On steep, north-facing acid volcanic rocks, such as the cliffs of Clogwyn Du'r Arddu, the yellow *Rhizocarpons* (of which *R. geographicum* is the commonest) are almost absent, and total lichen cover is quite low. Rock overhangs receiving moisture only by condensation provide a special niche for some crustose species, such as *Arthrorhaphis citrinella, Cystocoleus ebeneus, Opegrapha gyrocarpa* and *O. saxigena.* Most of these contain the alga *Trentepohlia* which, as a free-living species, demands shade. Some of these lichens exist as small dull crusts and are easily overlooked.

Rocks close to the soil remain moist for long periods and usually have a flora dominated by *Cladonia subcervicornis, Porpidia crustulata, P. macrocarpa, P. tuberculosa* and *Parmelia conspersa,* among others. The last named is often common in low lying areas where it forms pale green rosettes, with the lobes of the thallus contrasting with the large chestnut-brown apothecia. In the usually moist runnels, the characteristic species are the Rock Tripe (*Lasallia pustulata*), *Parmelia conspersa*, and the rich purple-brown rosettes of *Umbilicaria polyphylla.*

So far we have discussed the main components of the acid rock flora, but there are numerous minor variations. The Precambrian rhyolites of the Padarn ridge, for example, are often colonised by *Parmelia mougeotii*, yet this species is rare on the nearby Ordovician rhyolites. *Schaereria fuscocinerea* is characteristic of flinty rhyolite but much less common on the softer ignimbrite lavas of the same composition. Also, the striking quartz veins at Cwm Idwal and Crib Goch are often preferentially colonised by yellow *Rhizocarpon* species.

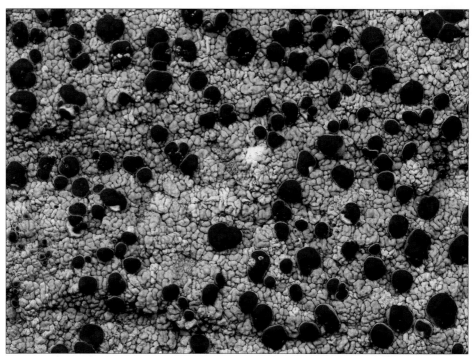

Ophioparma ventosa on rocks near Gallt yr Ogof.

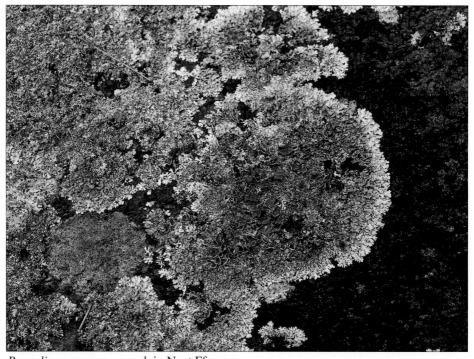

Parmelia conspersa on rock in Nant Ffrancon.

2. Neutral and basic volcanic rock.

Small exposures of dolerite and basalt are widespread in Snowdonia – good examples being those surrounding Llyn Teyrn, Snowdon and Cwm Cau on Cadair Idris. Although they contain a few percent calcium as orthoclase feldspar, the rocks weather slowly and often possess an acid-type lichen flora. However, these rocks tend to carry more species, and the presence of basic compounds can be inferred from the occurrence of *Ochrolechia parella* and *Tephromela* (formerly *Lecanora*) *atra*. Dark crusts of *Fuscidea* species are less evident, and pale crusts of species in the genus *Pertusaria* such as *P. corallina* and *P. lactea* are often seen. The andesite-basalts of Cwm Idwal possess a particularly rich flora, including *Amygdalaria pelobotryon* and *Rhizocarpon distinctum*. On the vertical faces other associations occur often including *Pertusaria flavicans, Toninia thiopsora* (*T. pulvinata*) and *Tylothallia biformigera*. The yellow *Rhizocarpon* species are all calcifuge, and rarely found on these rocks.

3. Slate and shale.

There are fewer natural exposures of these rocks as they are quickly reduced to small fragments by frost, but interesting communities are developing on old slate tips. Most of this rock lies within the Cambrian Slate Belt and is well exposed in the quarries at Bethesda, Llanberis and Blaenau Ffestiniog. It supports a calcifuge lichen flora with abundant *Fuscidea lygaea, Lecidea lactea* and yellow *Rhizocarpon* species. Due to rapid weathering, many exposures are colonised by pioneer species such as *Buellia aethalea, Placopsis gelida* and *Rhizocarpon lecanorinum*. Shaded slate has a similar flora to the lavas, but base-rich flushes are characterised by *Catillaria chalybeia, Haematomma ochroleucum, Lecanora gangaleoides* and *Lecanora* (*Lecidea*) *sulphurea*.

4. The bedded pyroclastic deposits

These are an interesting sequence of rocks consisting of layers of pumice-tuff, rhyolite lava and calcareous sediment. They are exposed in central Snowdonia, forming the cliffs of Gallt y Wenallt and Snowdon summit itself. These sites are usually well known because, in addition to lichens, they invariably have a rich bryophyte and vascular plant flora. The rock can be porous and is frequently vesicular, but this is just one of several factors that influences the richness of the lichen flora. Other factors including the amount of calcium present, and variations in the surface microtopography can also be important. The richest are on the more calcareous rocks, like those of Cwm Glas and Clogwyn y Garnedd on Snowdon – typical species being *Caloplaca citrina, Collema* spp., *Catapyrenium lachneum, Gyalecta jenensis, Psora decipiens* and *Trapeliopsis wallrothii*. On steep rock, *Lithographa tesserata, Pertusaria flavicans, Porpidia macrocarpa, P. speirea, Racodium rupestre* and *Stereocaulon vesuvianum* are often common, with *Fuscidea* species subordinate unless rhyolite inclusions are present. Flushes associated with these rocks may be rich in *Peltigera* species (dog lichens) including *P. apthosa, P. leucophlebia* and, more rarely, *P. venosa*. There are also several other rare or uncommon microlichens such as *Pannaria pezizoides*.

Stereocaulon vesuvianum on rock in Coedmawr.

5. Limestone

Pure limestones are rare in central Snowdonia, being occasionally inter-layered with the bedded pyroclastic deposits noted above. They have a distinct lichen flora with many species actually developing slightly embedded in the rock (endoliths). About 120 species have been recorded with many crustose species belonging to the genera *Aspicilia* (*A. calcarea, A. contorta*), *Caloplaca* (*C. aurantia, C. citrina, C. heppiana*) and *Verrucaria* (*V. calciseda, V. hochstetteri, V. nigrescens*).

HEATH AND MOORLAND

Several lichens are characteristic of heathland soils, often growing among bryophytes. Most of them belong to the large genus *Cladonia*. Common species of the Snowdonia heaths, and of British heathland in general, are *Cladonia arbuscula, C. diversa* (formerly *C. coccifera* agg.)*, C. ciliata, C. crispata* var. *cetrariiformis, C. furcata, C. portentosa, C. pyxidata* and *C. subulata*. Most of these species appear rapidly after heathland has been disturbed by cutting or burning, so in lichen terms they must be fast growing. Other heathland lichens include the shrubby dark brown *Coelocaulon aculeatum* and crustose species such as *Trapeliopsis pseudogranulosa* and *Baeomyces roseus*. Damp heathland also supports several large leafy lichens of the genus *Peltigera* such as *P. canina* and *P. lactucifolia*. The former was once powdered with pepper and consumed with warm milk as a supposed cure for rabies. In the more grassy areas *Peltigera membranacea* usually becomes the more common species.

Cladonia coccifera on Gyrn in the Carneddau.

Cladonia subulata with *Polytrichum* moss above Llyn Padarn.

Peltigera membranacea in Dyffryn Ogwen.

SPOIL TIPS

Mine waste is widely distributed in Snowdonia and can support interesting lichen communities. Waste rock from slate quarrying has already been mentioned, but there are also extensive wastes from a range of metal mining activities, many dating from the 18th and 19th centuries. This spoil is almost always highly acidic due to the presence of iron pyrites which is oxidised by bacteria and oxygen to sulphuric acid, and is generally toxic to plants. Surprisingly, such surfaces provide a habitat for several lichens. Common crustose species are *Acarospora sinopica, Rhizocarpon oederi* and *Tremolechia atrata*, all deep red in colour and containing large amounts of iron, and sometimes other metals. Foliose and fruticose species include *Placopsis gelida, Stereocaulon leucophaeopsis* and *S. vesuvianum*, all of which are associated with nitrogen-fixing algae. The inference is that they require such adaptations to compensate for nitrogen deficiency in the rock.

AQUATIC COMMUNITIES OF SEEPAGES, STREAMS AND LAKES

Seepages often posses a diverse lichen flora, and in common with the drier rocks summarised above, species richness increases with the calcium content of the rock and water. On acidic to neutral rock the lichen flora usually consists of *Ephebe lanata, Placopsis gelida, Porpidia macrocarpa, P. hydrophila* and *P.*

tuberculosa. Over base-enriched rock the floras often contain many nationally scarce or rare lichens. The common species are *Amygdalaria pelobotryon, Catillaria chalybeia, Lecidea phaeops, Porina lectissima* and *Stereocaulon vesuvianum.* More local are the foliose species *Solorina saccata* and *S. spongiosa*, which occasionally grow with the crustose *Staurothele fissa.* The richest sites for these lichens are in Cwm Idwal, Ysgolion Duon, Cwm Glas (Snowdon) and Cwm Meillionen (Beddgelert), but none has been studied exhaustively, and much remains to be discovered.

The rocky shores of mountain lakes provide habitats for some lichens. Good examples are the high cwms of Llyn Cau (Cadair Idris), Llyn Bochlwyd (Glyder Fach) and Llyn Ffynnon Lloer (Carneddau). A wide range of moisture regimes exist, and bird-perching sites are common on the small rocky lake islands. The littoral zones are usually dominated by the semi-aquatic lichen *Hymenelia lacustris,* but there are also some rarely recorded cyanolichens such as *Pyrenopsis granatina, Spilonema paradoxum* and *Thermutis velutina*, which merit further investigation. They form microscopic dark brown or blackish stains at lake margins where the rocks are subject to periodic wetting, and are easily overlooked.

The flora of permanent streams and rivers is normally richer than that of lake margins, probably because of differences in water chemistry. In a study of two adjacent mountain streams near Llanberis with contrasting water chemistry, large differences in the lichen flora were found. Both streams flowed over lava but the water draining over rhyolite lavas, and consequently more acidic (pH 6.3) and less calcareous, was notable for having much the poorer flora. Along with an abundance of blue-green algae, the main lichens here were *Ephebe lanata, Hymenelia lacustris* and *Rhizocarpon lavatum.* The adjacent base-enriched stream (pH 7.0) by comparison, had a richer flora, with *Catillaria chalybeia, Hymenelia lacustris* and *Thelidium papulare* among others. Water chemistry is clearly an important factor.

ARCTIC-ALPINE LICHENS IN SNOWDONIA

The land above 800 m has limited extent, but often supports elements of an arctic-alpine lichen flora. In common with other plant groups, some of these species could be a relict flora of the early postglacial period when they were presumably more widely distributed. Of the forty or so reasonably well-defined arctic-alpine lichen species of the Cairngorm mountains, about 60% occur in Snowdonia. However, only five (*Baeomyces placophyllus, Cetraria islandica, Cornicularia normoerica, Pseudephebe pubescens, Umbilicaria cylindrica*) are at all common, and many such as *Allantoparmelia alpicola* and *Thamnolia vermicularis* subsp. *subuliformis* are known from just a handful of sites. This suggests that a number of species are at the edge of their range under the present climate, and may be vulnerable to climatic warming. An interesting case currently under investigation is the distribution of *Thamnolia vermicularis* subsp. *subuliformis*. This fruticose species occurs in short turf on some of the highest summits, such as Cadair Idris. On Snowdon, it was moderately common on some of the highest slopes in the early 1970s but is now decreasing and becoming rare. The reason is not certain, but the area is subject to trampling and enrichment by sheep, so a climate change may not be the only factor. The whole region is also known to be affected by acid rain, to which many lichen species are known to be sensitive, along with other forms of air pollution. Even a small deterioration in air quality could therefore affect the flora. The effects are more likely to become evident at poorly buffered sites such as acid rock surfaces or the leached acidic soil and turf upon which *Thamnolia* grows.

The distribution of several other species is known in some detail. Two dark brown to black fruticose species, *Alectoria nigricans* and *Cornicularia normoerica* are practically confined to central Snowdonia at altitudes of 500 m or more. The former is typical of upland *Racomitrium lanuginosum* heath where it overgrows the moss, usually among frost-shattered rock. *Cornicularia normoerica* grows upon exposed acidic rocks and its dark colour may be an adaptation to high light intensity. This lichen is a calcifuge (acid-loving) and consequently absent from the summit rocks of Snowdon, but it is quite common on the Carneddau. It often grows with *Pseudephebe pubescens,* which looks like strands of brown wiry hair attached to the rock, and *Umbilicaria cylindrica,* which has a distinctive ash-grey thallus. Also associated with *Racomitrium* heath is the so-called Iceland Moss (*Cetraria islandica*), which is not a moss, but a lichen. Its distribution in the British Isles is unusual, being rare in the mountains of Eire which have an extreme Atlantic climate, but present on the heaths of lowland East Anglia, where the climate is more continental. It is interesting to note that the altitudinal limit of the species is depressed in western Snowdonia as compared with the east – possibly due to exposure to the strong moisture-laden winds coming off the Irish Sea. Iceland Moss is also distinguished by being one of the few edible lichens. Reindeer Moss (*Cladonia rangiferina*), also mis-named, is a lichen found in quantity only on the grassy slopes of the Carneddau. It is here that the best 'lichen heaths' occur, usually consisting of various *Cladonia* species. Extensive alpine heaths, such as those in parts of Scotland, do not exist in Snowdonia.

The close proximity of the sea may be a further reason for the relative scarcity of arctic-alpine lichens. Snowdonia experiences an Atlantic climate characterised by mild winters and wet cool summers. Most arctic-alpine regions are by contrast more 'continental', experiencing hot, dry summers and extremely cold winters. This difference may explain the absence of some well-known arctic-alpine species such as *Solorina crocata*, which grows on the central Scottish highlands. The remaining arctic-alpine species found in Snowdonia are all local in their distribution. They include several macrolichens growing directly on rock, such as *Allantoparmelia alpicola, Cetraria commixta, C. hepatizon* and *Umbilicaria proboscidea,* and some characteristic microlichens, such as *Lecidoma demissum, Rhizocarpon alpicola, Sporastatia polyspora* and *S. testudinea.*

CONSERVATION

The lichen flora of Snowdonia is rich, even for the west of Britain and contains a number of rarely recorded species. Local threats to the flora are fortunately few, but the lichens deserve monitoring for several reasons. The reduction in *Thamnolia vermicularis* on Snowdon, as previously noted, may be the result of higher grazing pressures and it is true that ground-inhabiting lichens are especially prone to soil disturbance. There are comparatively few terricolous (ground-dwelling) lichens, so the effect is unlikely to be catastrophic, and it is possible that the increased grazing is a consequence of recent milder winters leading to better plant growth on the peaks. On rocks, local damage has resulted from the damming of lakes for reservoirs, the altered water levels resulting in the loss of some rare species such as *Lecanora achariana* (a Red Data book species). While the effects of lake dams on invertebrate faunas are well known, the damage to lichen (and algal) communities has not been appreciated. Mountain rocks in general, are not damaged excessively by walkers or climbers so, although there is intensive 'boot' pressure in many parts of the National Park, lichens are little affected. The main worry concerns air pollution. Rain containing acidic pollutants from power stations and motor vehicles has been recorded in Snowdonia over several decades and quite a lot is now known about its effects on corticolous lichens. There is less knowledge though of the impact on saxicolous (or rock-inhabiting) species. However, the poorly buffered acidophile (acid-loving) communities that are so widespread in the area are likely to be most affected. Most of the recorded lichen losses from Snowdonia have been of species that grow on trees, and the felling of trees remains a significant threat. In the early 1980s the author witnessed the felling of a magnificent Ash by the roadside at Dolmelyllyn, which was covered with *Lobaria* species. Such losses are serious in an area where the *Lobarion* community has already been localised by a long history of old forest destruction and replanting with conifers. Other corticolous species for which we have old records but no recent ones include *Bryoria subcana*, *Collema nigrescens* (probably extinct), *Heterodermia obscurata*, *Imshaugia* (*Parmeliopsis*) *aleurites*, *Leptogium brebissonii*, *Opegrapha prosodea*, *Pseudocyphellaria norvegica* (probably extinct), *Strangospora pinicola* and *Usnea articulata*.

Despite the rich flora, there are few Red Data Book species in Snowdonia and it is perhaps significant that none of them is corticolous, despite the fact that in Britain, 59% of the RDB species grow on trees. The Snowdonia species usually grow over mosses, rocks or soil, and include *Bryoria smithii*, for which there are four records in Snowdonia occurring over mossy rocks in the lower hills; *Peltigera venosa*, a handsome species of calcareous soil, recorded from Snowdon but not seen for many years; and *Sticta canariensis*, a large lichen of mossy rocks in sheltered valleys which occurs sparingly in the Fairy Glen, Betws-y-Coed and elsewhere.

ADDITIONAL READING

Armstrong, R. A. 1974. The descriptive ecology of saxicolous lichens in an area of south Merionethshire, Wales. *Journal of Ecology*, 62: 33-46.

Kershaw, K. A. 1964. Preliminary observations on the distribution and ecology of epiphytic lichens in Wales. *Lichenologist*, 2: 263-276.

Pentecost, A. 1980. The lichens and bryophytes of rhyolite and pumice-tuf rock outcrops in Snowdonia, and some factors affecting their distribution. *Journal of Ecology*, 68:251-267.

Pentecost, A. 1987. The lichen flora of Gwynedd. *Lichenologist*, 19: 97-166.

Freshwater Algae

Allan Pentecost and Christine M. Happey-Wood

Introduction

It is now thought that the algae were the precursors of all other plants. The most primitive group, the blue-green algae (Cyanophyceae), are more closely related to the bacteria than to other forms of algae, and probably represent some of the oldest life forms. They are known as prokaryotic algae and lack, for example, the membrane-bound cell nucleus found in more advanced organisms. These primordial plants came into existence during the Precambrian period, possibly some 3 billion years ago, when life on Earth was in its infancy. At this period, multicellular organisms had yet to evolve, and the Earth's atmosphere was devoid of oxygen. However, by the start of the Palaeozoic era, some 600 million years ago, algae with a more complex cell structure had evolved. In addition to a membrane-bound nucleus, these 'eukaryotic' cells had the more elaborate internal structure which has become common to all life forms other than the blue-green algae and bacteria. During the early part of the Palaeozoic era, the eukaryotic algae radiated into a number of distinct lines, including the green algae (Chlorophyta), the red algae (Rhodophyta), and the brown algae (Phaeophyta), and some became multicellular. It is now thought that one of the ancient groups of grecn algae gave rise to all terrestrial plants.

From their apparent oceanic origins, algae have adapted to both terrestrial and freshwater habitats and now show a greater range of morphological and biochemical diversity than any other group of plants. Most of the terrestrial and fresh water forms are microscopic, occurring either as single cells, as simple or branched filaments or, more rarely, as multicellular thalli. The 1500 or so species found in Snowdonia fall into several groups or 'classes' such as the green algae (Chlorophyceae) or the golden brown algae (Chrysophyceae), distinguished by their modes of reproduction, their pigmentation and the chemistry of their storage products. The range of forms is enormous, from microscopic (but sometimes beautiful none the less) to macroscopic (visible without optical aid), and they are a fascinating group in many ways. Despite this, they are a neglected group, never attracting as much attention as lichens or bryophytes, for example.

Most freshwater algae belong to one of three classes: the blue-green algae – which are not well represented in Snowdonia, the green algae and the diatoms (Bacillariophyceae), and some representatives from the red algae (Rhodophyceae) and the golden brown algae. The diatoms are unique in having cell walls made of silica, and are consequently easily preserved and fossilised in lake sediments. This means that along with chrysophyte cysts (see below), they can provide insights into a lake's history. Although algae can be collected at almost any unshaded location, certain habitats tend to be especially rich in species. They include the water columns of lakes where species are held in suspension (phytoplankton), submerged vegetation, which provides attachment points for other species (periphyton), and the surface of soft sediments (epipelon). Vascular plants growing at shallow lake edges and bryophytes of permanent bogs can also be covered in a wide range of species. Other good sites include submerged rocks of rivers and streams, and the silt around lake margins. There are even 'aerophyte' algae that live on exposed rocks and soil.

THE HISTORY OF ALGAL STUDIES IN SNOWDONIA

The earliest published records of freshwater algae in Snowdonia are those of John Jacob Dillenius in his *Historia Muscorum* (1742). Very little is known about his collecting but, according to Woodhead and Tweed (see below), he found the red alga *Lemanea fluviatilis* in the Afon Cegin. In his *Synopsis of the British Confervae* (1809), Lewis W. Dillwyn included a few filamentous species such as *Scytonema* (*Conferva*) *myochrous* collected by Dawson Turner from Beddgelert in 1801, and *Batrachospermum moniliforme* from Cwm Glas, Snowdon. Although concentrating mainly on Ynys Môn, the Rev. Hugh Davies included a few records for Snowdonia in his *Welsh Botanology* (1813), and J. E. Griffith also recorded some algae from the region in his *Flora of Anglesey and Carnarvonshire* (1895), including *Rivularia biasolettiana* from Bangor where it still occurs. The algologist J. Roy visited Snowdonia in the mid 19th century but published nothing on his collections. John Ralfs collected at Caernarfon in 1841, but his most important work was done at Dolgellau. His records were published in the *British Desmideae* (1848) but, although it was regarded as a seminal work in its time, it was eventually to be overshadowed by West and West's *Monograph of the British Desmidiaceae* (1904-1923) (see below). Collections of desmids, a distinctive group of green algae (see below), were also made near Capel Curig by A. W. Wills, and he published two small articles on his finds in the *Midland Naturalist* (1881). He was perhaps the first to draw attention to the richness of the area for algae. While such records are interesting historically, most of our knowledge comes from two pairs of collectors: the father and son team of William and George West, and the partnership of Norman Woodhead (a teacher at the University College of North Wales) with Ronald Duncan Tweed (a wealthy amateur botanist). The Wests made a number of visits in the late 19th century, and published their findings in the *Journal of the Royal Microscopical Society* (1890). They listed more than one hundred species, most of them diatoms and desmids from Snowdon, Llyn Idwal and Capel Curig. In their great work: *The Monograph of British Desmidiaceae* (1904-1923), they gave a detailed account of the desmids for North Wales. The Wests also sampled at least 19 of the North

Wales lakes, and recorded 126 species of which 101 were desmids. The results of this survey appear in their account of the British freshwater phytoplankton published as part of the *Proceedings of the Royal Society* (1909). Most of their locations are imprecise, although their original collections still survive at the Natural History Museum, along with George West's notebooks. From their collections it can be deduced that at least nine visits were made to the Snowdon area between 1880 and 1912. According to Woodhead and Tweed (see below) over 400 species of desmids have been found in North Wales.

Other records were obtained from Llyn Ogwen by B. M. Griffiths, who compared the flora with that of some Ynys Môn (Anglesey) lakes. His results were published in the *Journal of the Linnean Society* (1926). However, his study has now been overshadowed by the work of Woodhead and Tweed who began their investigations in the 1930s and continued into the mid 1950s, publishing their results in *Northwestern Naturalist* (1954-5). They examined over 1500 collections from Anglesey and Caernarvonshire, and were the first algologists to systematically measure water pH (acidity) at their collecting sites. Their collections were left to the University College Bangor herbarium but recent enquiries have failed to locate them. They recognised about 1500 species, with the flora of Cwm Idwal alone exceeding 600 species, making it one of the richest algal sites known on Earth.

In the latter part of the 20th century few species have been added to the flora, but the algae have continued to be the subject of various research programmes. These have tended to concentrate on specific locations and to focus on algal ecology, growth dynamics and primary production. During the 1970s an interesting investigation of the species found on rock surfaces of Cwm Idwal was carried out by Tim Allen, then a research student at the School of Plant Sciences, Bangor. He investigated rock surfaces at Cwm Idwal and Bangor and found that slope and aspect influenced the algal flora This was followed in June 1973 by a survey by M. J. Liddle (and others) of the phytoplankton of twelve Snowdonia lakes, ten within Coed Gwydir and the two Llynnau Mymbyr (Mymbyr lakes).

Two mountain lakes, Llyn Padarn and Llyn Peris, have been studied in detail by various scientists from the School of Plant Sciences including Dr J. Priddle and the present two authors. In the period from 1970 to the present these lakes have contained contrasting phytoplankton floras due to the input of treated sewage into Llyn Padarn, and as a result of mining for copper and slate around Llyn Peris. Thirty years ago the phytoplankton of Padarn, the lower lake of the two, was dominated by centric diatoms, particularly *Cyclotella glomerata*, but *Asterionella formosa* then became the dominant species, to be superseded in the early 1970s by *Fragilaria crotonensis*. Major algae in the phytoplankton of Llyn Peris during this period were *Dinobryon,* and *Rhodomonas* species, other small chlorophycean and chrysophycean flagellates and *Tabellaria flocculosa*. The damming of Llyn Peris as part of the Dinorwic hydroelectric scheme in the late 1970s – separating it from Llyn Padarn – has meant that both lakes have now been altered by human activity. As a consequence of nutrient enrichment, the lower lake (Padarn) has gained algal biomass, but lost some of its species diversity in the process. The upper lake (Peris), with heavy metal contamination and frequent changes in water level, has similarly lost diversity, but in this case with no gain in biomass.

Llyn Padarn looking towards Snowdon.

THE ALGAL FLORA OF SNOWDONIA BASED ON RECENT STUDIES

As we have already mentioned, the larger colonial and filamentous blue-green algae are poorly represented in Snowdonia, and there are at least two possible explanations. First that the geology of the area is deficient in the base-rich materials which are believed to increase species richness, although exactly why this should be so is unclear. And second that nutrient levels in these waters are simply too low. Blue-green algae are well known for the formation of dense growths (water-blooms) in lakes where the concentration of nutrients become high. This can occur in lowland lakes due to natural bacterial processes, but is often exacerbated by chemical run-off from farmland and inputs of treated sewage. The scarcity of the bloom-forming algae *Anabaena*, *Aphanizomenon* and *Microcystis* shows that with few exceptions (Llyn Padarn is a notable one) Snowdonia's waters are very pure. However, not all blue-green algae are associated with enrichment, and species inhabiting damp rocks in the mountains, such as *Gleocapsa*, are common. A few unusual and rarely recorded species include *Crinalium endophyticum*, found at Fairy Glen, near Betws-y-Coed in the 1920s. Recently though, with improvements in microscopic techniques, a number of minute species of blue-green algae of the genera *Synechococcus* and *Synechocystis* have been described from the phytoplankton of lakes in Snowdonia. On occasions in late summer, these tiny components of the phytoplankton (picoplankton) have been found in their millions in Llyn Padarn as well as in lesser quantity in other aquatic habitats such as water-filled slate quarries, bog pools, lakes and rivers. The sections that follow are concerned with a more detailed discussion of the other main algal groups.

THE GREEN ALGAE (CHLOROPHYTA)

The Volvocales are a large group with many species capable of swimming using a whip-like tail or flagellum. They range from minute unicellular species to larger colonial ones which may be visible to the naked eye. Many of the unicellular species (Chlamydomonads) are common components of the phytoplankton and sediment, as well as being associated with macrophytes, of nutrient-poor (oligotrophic) waters. They have been found in a number of Snowdonia lakes such as Llyn Cwellyn and Llyn Llydaw. The larger colonial Volvocales such as *Eudorina*, *Pandorina* and *Volvox* are poorly represented, and this probably reflects their requirement for the nutrient-rich waters associated mainly with the lowlands. However, the colonial *Paulschulzia* has been common in the phytoplankton of Llyn Padarn in recent years, and there has been a rise in *Eudorina* and *Pandorina* species. Numbers of species belonging to the large order Chlorococcales also seem to have increased. They live suspended in water like the Volvocales, but differ in their lack of swimming appendages. The small *Chlorella*, *Ankistrodesmus falcatus*, *Scenedesmus graheinsii* and needle-shaped *Monoraphidium*, a few microns in size may occur in large numbers in the plankton of lakes such as Llyn Padarn, Cwellyn and the extremely nutrient-poor

Llyn Llydaw. The high population densities of these minute algae may be attributed to their small size, since it is this that gives them their high surface area to volume ratio, and makes them efficient as nutrient absorbers. Other chlorococcalean algae such as *Scenedesmus* are more common in lowland situations and often occur at sites along the coastal fringe of Snowdonia – on the duneland south of Caernarfon for example, or in the Ynys Môn (Anglesey) lakes. A species of *Chlorella* which lives symbiotically within the ciliate *Ophyridium* sometimes forms large jelly-like masses at the edges of mountain lakes, though little is known about its biology or ecology. Green algae can live symbiotically with many other organisms – with various protozoa for example, particularly in acidic sites, and with freshwater sponges, turning them green. Perhaps their best-known symbiotic relationship, though, is the one they form with fungi in those supremely hardy organisms, the lichens.

Two other large groups of green algae well represented in Snowdonia are the Zygnemales and the Desmidiales. Both are known as 'conjugating' algae characterised by a method of sexual reproduction in which fusion of male and female cells takes place without the involvement of swimming sperm cells. The Zygnemales consist mainly of filamentous species, the most familiar being *Spirogyra*. However, the related genus *Mougeotia* is more common in Snowdonia and occurs throughout the region, often in abundance. Few species have been recorded because they can be identified only when they are undergoing sexual reproduction – a rarely observed phenomenon in montane Britain. Non-filamentous members of the Zygnemales, the saccoderm desmids, which have a simple cell wall structure, are widespread. They grow profusely among mosses in the bogs of the central Snowdonia cwms. Common species here are *Netrium digitus* and *N. oblongum*, but there are also rarities such as *Genicularia elegans*, *Roya cambrica* and several species of *Spirotaenia*. The Placoderm ('true') desmids are characterised by their complex, often-ornamented cell walls, and their bilateral symmetry, with a central 'waist'. Many are beautiful microscopic objects and have their own dedicated group of desmidology specialists. William and George West were devoted to this group, describing many new species from Snowdonia. Among the most widespread and common species in the area are *Closterium acerosum*, *C. parvulum*, *C. striolatum*, *Cosmarium brebissonii*, *C. cucumis*, *C. margaritiferum*, *C. tinctum*, *Euastrum binale*, *E. elegans*, *Hyalotheca dissiliens*, *Staurastrum margaritaceum*, *S. paradoxum*, *S. punctulatum* and *Tetmemorus granulatus* – found in shallow boggy pools, among aquatic vegetation. Typical montane species mainly restricted to the higher cwms included *Closterium didymotocum*, *Cosmarium annulatum*, *C. parvulum*, *C. reniforme*, *Euastrum cuneatum*, *E. pectinatum*, *Micrasterias denticulata*, *M. oscitans*, *M. truncata* and *Xanthidium armatum*. The region undoubtedly has one of richest desmid floras in Britain, a fact well recognised by the Wests, and prompting them to declare in their 1909 paper that 'the Capel Curig lakes were… superior to any known in the world for their desmid flora to date…' According to the labels on G. S. West's original collections, the area they covered was probably to the north, perhaps in the neighbourhood of Llyn Crafnant. This region, and a few others, notably Cwm Idwal, provide sanctuary for several rare desmids including *Docidium baculum* and *Onychonema filiforme*.

Previous studies have shown that certain desmids fluctuate greatly in abundance from year to year, and perhaps even between decades. Such variability is so far unexplained, but it seems likely that rainfall amounts and perhaps temperature are involved. The large desmid *Staurastrum planktonicum*, for example, bloomed in Llyn Padarn in 1992, turning the lake and out-flowing river bright green. Not surprisingly, this attracted much public attention. In 1992, low rainfall in early summer meant that water flow through the lake was much reduced. A more predictable phenomenon is the seasonal variation in the abundance of planktonic algae. In deep lakes this is caused by the changing properties of the water column (stable in summer, unstable in winter), which influences the availability of nutrients. This is because in summer when winds are light, the upper layers of a lake warm up making the water lighter and unable to sink below the colder and darker water below. In the upper layer, nutrients are removed by the growing algae, leading to nutrient depletion near the surface. Levels of irradiance and temperature also play their part, some species showing a preference for warm water, others for cold. As a result, groups such as the diatoms, which can cope with an unstable water column, are common in spring and autumn, while many greens develop in the more stable summer water. The situation is somewhat different and largely unexplained in bogs and pools and may be more dependent upon water availability and growth of higher plants. Detailed monitoring of key sites is required to clarify these phenomena.

Among the larger (macroscopic) species of green algae are a variety of filamentous forms. Conspicuous populations of the simple (unbranched) *Ulothrix* (e.g. *U. zonata*) and branched *Chaetophora* (e.g. *C. incrassata*), *Draparnaldia glomerata* and *Stigeoclonium tenue* grow attached to stones in many of the streams and rivers of the region. They are bright green and often conspicuous in running water throughout the region during summer.

THE RED ALGAE (RHODOPHYTA)

Although there are relatively few 'non-marine' red algae, those found in Snowdonia are all fairly common, especially in the higher altitude streams and rivers. *Hildenbrandtia rivularis*, for example, found as a reddish-purple crust on stones in the streambeds – especially where the streams are shaded as in some of the deeper wooded valleys. *Lemanea fluviatilis*, with its stiff olive-green bristle-like appearance and swellings along the length of its filaments, grows on stones and rocks in fast flowing and even torrential waters such as those of the Afon Ogwen. The elegantly branched, filamentous *Batrachospermum moniliforme* is found in calmer water. It occurs in the upper reaches of slower flowing streams and in shallow lakes where it often appears greenish rather than pink-red because of bleaching by sunlight. It occurs in quantity, for example, under the shady banks of Llyn Teyrn on the slopes of Snowdon.

THE DIATOMS (BACILLARIOPHYTA)

The diatoms are a particularly fascinating group. They are unicellular with each cell consisting of two valves (termed frustules) made of polymerised silica, and often sculpted into intricate and beautiful patterns. They are broadly classified into boat-shaped 'pennate' types, and disk-shaped 'centric' types. Phytoplankton contains many diatom species, separable into those typical of lowland waters and those found commonly in the uplands. In lowland lakes, such as Llyn Padarn, common diatoms include *Asterionella formosa*, *Diatoma vulgare*, *Fragilaria crotonensis* and *Melosira varians*. In mountain lakes and corries such as Llyn Ogwen these species are all uncommon, but others, such as *Fragilaria capucina*, *F. virescens*, *Tabellaria fenestrata* and *T. flocculosa* become much more abundant. The most diverse diatom floras, predominantly pennate types, are found among higher plants and on sediments in shallow lakes and watercourses. These too may be divided broadly into the species occupying lowland lakes with moderately high nutrient levels (mesotrophic species), and those characteristic of upland low-nutrient waters (oligotrophic species). Common lowland species include *Cymatopleura solea*, *Gyrosigma* spp., *Navicula cryptocephala* and *N. radiosa*. Montane species of lakes such as Llyn Idwal include *Diatoma hiemale*, *Eunotia arcus*, *E. exigua*, *Frustulia rhomboides*, *Navicula pupula*, *Neidium iridis*, *Peronia fibula* and *Pinnularia* spp. However, some of the commonest diatoms, such as *Achnanthes minutissima*, *Cocconeis placentula*, *Cymbella cistula* and *Meridion circulare*, appear to be indifferent to altitude. At the other extreme, many species have rather exacting requirements, and can be affected by quite small variations in water chemistry – such as reduced acidity. The type of substratum can also be an important. *Achnanthes lanceolata* and *Cymbella cymbiformis* are found more frequently in base-enriched water – on the bedded pyroclastic deposits of Snowdon, for example – while *Tetracylus lacustris* is one of several species associated with mosses attached to damp rocks.

ARCTIC AND ALPINE FLORAS

A careful investigation of some high altitude sites by N. Woodhead and R. D. Tweed led to their 1947 paper 'Some algal floras of high altitudes in Snowdonia' in the *North Western Naturalist*. They studied, among other waters, Ffynnon Oer (at 975 m), the highest spring on Snowdon and Ffynnon Llyffant on Carnedd Llewelyn, at 830 m, regarded by some as the highest tarn in England and Wales. They found the algal flora of the latter disappointing, although they did collect the uncommon blue-green alga *Eucapsis alpina* there. Altogether, 328 algal species were recorded from these upland springs and pools, though comparatively few can be considered truly alpine. On Snowdon, they examined two small pools below Y Lliwedd, one dominated by sedges and the other by *Sphagnum* moss. They were only a few metres apart, yet had strikingly different algal floras, emphasising the importance of the higher vegetation as a controlling influence. Certain genera (*Cymbella, Euastrum, Gomphonema, Navicula,*

Pinnularia) were also noted as absent or rare from the aquatic habitats of exposed high ground. Species recognised as arctic-alpine were found, but they were usually rare or sporadic. Among the desmids were *Cosmarium costatum, C. decedens, C. etchachensis, C. ochthodes, C. reniforme, C. elevatum* and *Euastrum montanum* (considered to be a glacial relict species). Arctic alpine diatoms included *Caloneis alpestris, C. latiuscula, C. obtusa, Cymbella cesati* and *Neidium bisulcatum*. Another alga of mountain streams, *Hydrurus foetidus*, regarded as sub-alpine, was also noted, but only in Llyn Ogwen. Woodhead and Tweed's sampling sites were small in area but some of their locations were well documented, providing a rare opportunity to resample from precisely known locations.

POSTGLACIAL HISTORY – DIATOMS AND CHRYSOPHYTES

The diatoms are characterised by their box-like silica cell walls, and many of the flagellates (cells with whip-like tails) are covered by silica scales and produce silica-walled cysts during reproduction to tide the cells over unfavourable periods. When golden-brown (chrysophyte) algae and diatoms die, their silica cells and cysts sink to the bottom of lakes and accumulate in the sediments. They may be preserved for long periods, so that the history of a lake and its catchment in terms of its algal flora can often be deduced from study of the bottom sediments. If such data are linked to radiometric dating of the sediments and to ecological data, it becomes possible to follow climatic and other changes through time, and to date them.

Such studies have been carried out on several lakes and bogs including Llyn Llewesig, Cors Geuallt (a valley bog near Capel Curig), two small oligotrophic mountain tarns, Llyn Clyd and Llyn Glas, and the Llanberis lakes, Llyn Padarn and Llyn Peris. The sediment cores from Llyn Padarn dated back over 6000 years but only the last thousand years in Llyn Peris.

It appears that the waters in the Snowdonia lakes had a higher mineral content in the few thousand years of the immediate postglacial period. Diatoms typical of alkaline conditions were found in the oldest sections of cores. In Llyn Clyd and Llyn Glas, *Fragilaria brevistriata* accounted for most of the diatom fossils at the base of the cores. *Fragilaria construens, F. pinnata* and *F. elliptica* were more numerous in the early postglacial period together with *Achnanthes levanderi, A. minutissima, A. cryptocephala, Amphora ovalis, A. pediculus, Navicula seminulum, N. frustulum* and other alkaliphilous (alkaline-loving) taxa. During the Atlantic Period (7000-6000 years ago) Llyn Clyd was dominated by *Cyclotella comensis* while *Tabellaria flocculosa* and *Fragilaria* had declined, indicating that it had become more acidic. Coincidently, in Llyn Glas, the typically acid-loving *Achnanthes minutissima* var. *cryptocephala* was the major species. The lake also supported populations of *Eunotia* and *Frustulia*, which prefer more acidic conditions. It can be surmised, therefore, that by this time, most of the lakes in Snowdonia had become acidic. For Llyn Clyd, a third diatom zone corresponding to the late Flandrian (c.5000-6000 years ago) was recognised in which the dominant species were *Achnanthes minutissima* var. *cryptocephala* with *Anomoneis exilis* var. *lanceolata, Synedra rumpens, Eunotia* spp. and *Tabellaria flocculosa* in higher frequencies than for the lower two thirds of the core. In the oldest sediments from Llyn Padarn (6000 years ago), the centric diatom *Melosira distans* was dominant with small numbers of *Cyclotella* species. Also at this time, the numbers of siliceous chrysophyte cysts incorporated into the sediments of Llyn Padarn were at a maximum. These included *cysta minima* and *cysta microspinosa* and confirm the acid peaty conditions of the lake water at the time. From 4500 years ago, *Melosira distans* was still dominant in the sediment profile but, as time went by, it was gradually replaced by three species of *Cyclotella* including *C. ocellata*, together with *Tabellaria flocculosa, Achnanthes* sp. and *Anomoneis exilis*. These last three diatoms were the major taxa from 3300 to 2700 years ago during the period of deciduous forest when the water had become slightly richer in nutrients and a little less acidic.

The diatom populations in the lakes in Snowdonia have therefore changed markedly since the immediate postglacial period, from those dominated by species characteristic of alkaline water to those characteristic of oligotrophic, peaty acidic waters. Altitudinal differences also had an influence, however. Llyn Padarn was dominated by the acidophilous *Melosira distans* by 6000 years ago, while in the higher mountain tarns this diatom did not appear until much later and in smaller proportions. Similarly in the tarns, the change from mineral-rich conditions occurred later than in Llyn Padarn. At this time in Llyn Glas and Llyn Clyd, the major algae were more intermediate in pH preference, and species of alkali-loving *Fragilaria* were still present in Llyn Glas. There were differences between the two tarns, however, and changes appear to have been less in Llyn Glas and to have taken place more slowly. Typically planktonic species such as *Tabellaria flocculosa* never established here as they did in Llyn Clyd.

During more recent times, the diatoms preserved in the sediment of some Snowdonian lake cores begin to show the influence of man and his activities. In Llyn Padarn, *Asterionella formosa*, a species indicative of nutrient rich (eutrophic) waters became dominant about 2000 years ago. There was then an increase in the sodium and potassium content of the sediment core, indicating that much of the catchment had been stripped of the protection from leaching conferred by a covering of vegetation. It seems likely, therefore, that this period was marked by disturbance and agricultural development. Although this would have coincided with the period of Roman occupation, their presence seems to have been of little consequence to the Llanberis area apart from their need for grain possibly stimulating agricultural development. Following this period of enrichment, Llyn Padarn reverted to a more nutrient-poor (oligotrophic) lake from about 1900 years ago. This was reflected in its diatom flora, which came to resemble that of the period from about 3300 years ago until the time of the disturbance. From about 1250 years ago the alga populations became more diverse, with *Cyclotella kuetzingiana, Skeletonema subsalum, Fragilaria construens* var. *subsalum* and several pennate diatoms, particularly *Navicula* together with other species of *Cyclotella, Tabellaria, Achnanthes* and *Anomoneis*. By this time deforestation was complete, and the catchment area was dominated by grass and heathland plants. This brought about a decrease in lake nutrient levels, allowing the diverse diatom flora characteristic of more oligotrophic water to develop.

A sediment core from Llyn Peris, which until the construction of the Dinorwic pump storage hydroelectric scheme in the mid-1970s was linked to Llyn Padarn, spans a shorter period of history. The diatom assemblage from the sediments of 1000 years ago proved to be comparable with that of Llyn Padarn from 1900 years ago. Around 500 years ago, the centric diatoms *Cyclotella* and *Skeletonema* in Llyn Peris declined dramatically. These dominants were succeeded by species such as *Rhizosolenia eriensis* and *Synedra rumpens*, while the numbers of *Tabellaria flocculosa*, several species of *Achnanthes, Anomoneis exilis*, and other pennate diatoms including *Navicula cryptocephala, N. minima* and *N. seminulum* increased. This abrupt change in the diatoms corresponded to a high copper concentration in the sediment core – a consequence of mining in the Llanberis Pass. It is interesting to note that many of the shells (frustules) of *Synedra rumpens* were distorted and bent in several places, possibly as a response to this pollution. Also in Snowdonia, *Rhizosolenia eriensis* appears to be unique to the sediment profile of Llyn Peris. Thus the algal flora of Llyn Peris from 500 years ago until about 100 years ago, during the period of active copper and slate mining, was very different from that of other lakes in Snowdonia. *Rhizosolenia eriensis* is no longer present in Llyn Peris, and was not found in the surface regions of the sediment core. It probably disappeared therefore during the later part of the 19th century. Other diatoms present in Llyn Peris during this period and subsequently lost include a small assemblage of the centric species *Melosira distans, Cyclotella kuetzingiana, C. ocellata*, other species of *Cyclotella, Skeletonema subsalum* and two large populations of *Tabellaria flocculosa* and *Anomoneis exilis*. Since the early part of the 20th century, centric species have not been found, but the pennate diatoms have become more diverse as the contributions of *Tabellaria flocculosa* and *Anomoneis exilis* have declined. The assemblages of chrysophyte cysts found in the sediments of Llyn Padarn and Llyn Peris were similar until mining began. But, at the end of the 19th century, the numbers from Llyn Peris were about double those from Llyn Padarn, and the types more diverse. Only in the past 50 years have the chrysophyte cysts from both lakes declined.

The history of a variety of upland lakes in North Wales from about 1800 onwards has been investigated from short sediment cores by Rick Batterbee and his colleagues in the Paleoecology Research Unit, University College, London. In their studies of the Gyno, Hir, Cwm Mynach, Dulyn, Eiddew Bach, Llagi, Gammallt, and Conwy lakes, and in the lakes of the Berwyn, they found floras of fossilised diatoms that were typical of those in lakes with increasing acidity. But in the past two hundred years, the species have changed from those characteristic of waters with a pH of around 6.0, to those of more acid conditions with pHs as low as 4.5 (Llyn Llagi) and 5.6 (Llyn Cwm Mynach). In Llyn Llagi, south of Snowdon and the Afon Glaslyn, the sediments were dominated by attached neutral and acid-loving species. Prior to 1850, the dominant diatoms were *Eunotia veneris, Anomoneis vitrea* and *Achnanthes minutissima*, and the lake had a pH of between 5.9 and 6.2. *Achnanthes minutissima* began to decline in the late 19th century, but the other two diatoms have decreased in importance only during the second half of the 20th century. From about the mid-19th century, *Eunotia lunaris* and *Frustulia rhomboides* were found, and by the turn of the century *Achnanthes marginulata* had been recorded. Around 1960, *Tabellaria quadriseptata* appeared in the cores and, by 1985, it had become the dominant diatom. *Navicula heimansii* and *Melosira lirata*, not previously recorded there, were also found. Thus the general trend of acidification in surface waters in Snowdonia has resulted in significant changes in the diatom populations, showing how sensitive these lakes are to environmental change.

Palaeoecological investigations at Llyn Tegid (Bala Lake), the largest natural lake in Wales, have demonstrated changes in the diatom populations rather different to those in the upland lakes. In the 1970s, the sediment profiles of diatoms derived from the phytoplankton were dominated by *Cyclotella glomerata* with *Aulacosira subarctica, Achnanthes minutissima, Cyclotella comensis, C. stelligera, Fragilaria capucina* and *Tabellaria flocculosa* in decreasing order of importance. By 1996 *Asterionella formosa* was the major species found in the sediment, with lower proportions of *Cyclotella glomerata, C. stelligera, C. comensis* and *Tabellaria flocculosa*. Additionally, *Fragilaria crotonensis* was present in 1994, but with considerably lower numbers in 1996. It is interesting that these changes in the planktonic diatoms in Llyn Tegid during the last twenty years mirror those for Llyn Padarn since the late 1960s, with *Cyclotella glomerata* being replaced by *Asterionella formosa*, clearly showing the impact of nutrient enrichment from the fertilisers used on surrounding agricultural land.

ADDITIONAL READING

Belcher, H. & Swale, E. 1976. *A beginner's guide to Freshwater Algae*. Institute of Terrestrial Ecology: HMSO.

Lind, E. M. & Brook, A. J. 1980. *Desmids of the English Lake District*. Freshwater Biological Association Scientific Publication no. 42.

Pentecost, A. 1984. *Introduction to Freshwater Algae*. Richmond Publishing, Surrey.

Van den Hoek, C., Mann, D. G. & Jahns, H. M. 1995. Algae. *An Introduction to Phycology*. Cambridge University Press.

NOTES ON CONTRIBUTORS

CHARLES ARON works as a freelance naturalist specialising in mycology. For much of the past ten years he has been carrying out a detailed study of the fungi of Anglesey, Caernarfon and Meirionnydd. He has recorded many species new to Wales and aims to publish a detailed fungus flora of north-west Wales in the near future.

TIM BLACKSTOCK has worked in North Wales since 1979, firstly for the Nature Conservancy Council (NCC) and then the Countryside Council for Wales (CCW). His professional expertise covers the ecology and conservation of upland and lowland habitats across Wales, and he has a particular interest in the composition of the Welsh bryophyte flora. He has been actively engaged in recording the distribution of mosses and liverworts in north-west Wales for over twenty years.

DAVID EVANS is recently retired after a career that began in electronics, but switched to nature conservation at the age of 30. He joined the Nature Conservancy in Bangor in 1970, and continued there in its various incarnations for the next 30 years. His work was mainly in botanical survey and monitoring particularly in Snowdonia, and included a pioneering survey of Limestone Pavement in Great Britain at a time when the threats to its survival were not generally appreciated. Photography has also been a lifelong interest.

CHRISTINE HAPPEY-WOOD gained a PhD in algal ecology at Bristol University. She was then appointed to the staff of the School of Plant Biology (now part of Biological Sciences) at the University of Wales, Bangor, where her research included the ecology of algae in the lakes of Snowdonia. Although taking early retirement recently to give her more time for research, she continues her studies of algae, and is currently doing algal research with the Environment Agency.

BARBARA JONES has had a long-standing interest in many aspects of mountains, including their geomorphology, ecology and vegetation, and has spent many a happy hour climbing and mountaineering in Snowdonia and other parts of the world. Since 1985, she has worked for the NCC and CCW in Scotland and North Wales, recently completing a PhD on the ecology of the Snowdon Lily, which further developed her interest and specialism in arctic-alpine cliff and montane flora.

DEWI JONES is a local amateur botanist mainly interested in the arctic-alpine flora of Snowdonia. He is a keen hill walker and searcher for rare plants, and has combined this with original studies on the history of botanical exploration in the region. He has published several papers on the subject, and is the author of *The Botanists and Guides of Snowdonia* (1996).

ALLAN PENTECOST first became interested in freshwater algae when he was just eight years old. Some years later, after graduating in botany at Imperial College London, he moved to the Marine Science Laboratories, Menai Bridge, where he completed a PhD on the freshwater algae of North Wales. During this period he also developed an interest in lichens and published a lichen flora of Gwynedd. He is now a lecturer at the University of London, but keeps up his interest in the lichens and algae of Snowdonia.

PETER RHIND has worked as an ecologist for the Field Studies Council (FSC) and the Countryside Council for Wales (CCW), where he now specialises in coastal ecology. He has a PhD in marine biology, but has always maintained a keen interest in terrestrial botany. Since joining CCW's Bangor Office in 1992, he has developed a special interest in the plants of Snowdonia, and now also specialises in plant photography and digital imagery. His many publications include a vascular plant flora of Great Cumbrae Island (Scotland) and a bryophyte flora of Pembrokeshire.

MARCUS YEO has recently moved to the Joint Nature Conservation Committee in Peterborough, having previously worked for the NCC and CCW in Bangor. He specialised in scientific aspects of upland conservation and developed a particular interest in the plant communities of Welsh hills and moorlands. He has a long-standing interest in bryophytes and, over the years, has acquired a detailed knowledge of the bryophyte flora of Snowdonia, with several new species records to his credit.

ILLUSTRATION CREDITS

P. ii (Map of the Snowdonia area) A. Bradley; p. xiii (Cwm Idwal) P. Rhind; p. 3 (Thrift) P. Rhind; p. 4 (Mountain Sorrel) D. Evans; p. 5 (Mossy Saxifrage) D. Evans; p. 8 (Castell y Gwynt) P. Rhind; p. 9 (Welsh Poppy) P. Rhind; p. 10 (Roseroot) P. Rhind; p. 11 (Tufted Saxifrage) D. Evans; p. 13 (Snowdon Lily) P. Rhind; p. 14 (Purple Saxifrage) P. Rhind; p. 15 (Starry Saxifrage) D. Evans; p. 15 (Alpine Saxifrage) P. Rhind; p. 16 (Moss Campion) P. Rhind; p. 17 (Alpine Saw-wort) P. Rhind; p. 17 (Mountain Avens) P. Rhind; p. 21 (Bog Bean) P. Rhind; p. 22 (Cotton-grass) D. Evans; p. 23 (Round-leaved Sundew) P. Rhind; p. 24 (Common Butterwort) D. Evans; p. 25 (Heathland at Cilgwyn) P. Rhind; p. 26 (Bog Asphodel) D. Evans; p. 27 (Juniper) P. Rhind; p. 30 (Nantgwynant woodland) P. Rhind; p. 31 (Wood-sorrel) P. Rhind; p. 32 (Globe-flower) D. Evans; p. 39 (Forked Spleenwort) P. Rhind; p. 41 (Tunbridge Filmy-fern) P. Rhind; p. 44 (Parsley Fern) D. Evans; p. 44 (Beech Fern) P. Rhind; p. 45 (Green Spleenwort) P. Rhind; p. 45 (Holly-fern) P. Rhind; p. 46 (Alpine Woodsia) P. Rhind; p. 46 (Oblong Woodsia) P. Rhind; p. 47 (Oak Fern) D. Evans; p. 47 (Polypody) D. Evans; p. 48 (Lemon-scented Fern) P. Rhind; p. 50 (Afon Rhaeadr) P. Rhind; p. 51 (Royal Fern) P. Rhind; p. 54 (Bracken in Nant Ffrancon) P. Rhind; p. 60 (Alpine Clubmoss) P. Rhind; p. 60 (Fir Clubmoss) P. Rhind; p. 61 (Stag's-horn Clubmoss) P. Rhind; p. 62 (Lesser Clubmoss) P. Rhind; p. 62 (Marsh Clubmoss) P. Rhind; p. 67 (Wood Horsetail) P. Rhind; p. 67 (Marsh Horsetail) P. Rhind; p. 68 (Great Horsetail) P. Rhind; p. 71 (*Bryum alpinum*) P. Rhind; p. 72 (*Breutelia chrysocoma*) D. Evans; p. 76 (*Gymnomitrion obtusum*) P. Rhind; p. 77 (*Oedipodium griffithianum*) P. Rhind; p. 77 (*Pterogonium gracile*) P. Rhind; p. 78 (*Polytrichum piliferum*) P. Rhind; p. 79 (*Racomitrium lanuginosum / R. fasciculare*) P. Rhind; p. 80 (*Hylocomium splendens*) P. Rhind; p. 80 (*Leucobryum glaucum*) P. Rhind; p. 81 (*Thuidium tamariscinum*) P. Rhind; p. 81 (*Lepidozia cupressina*) P. Rhind; p. 82 (*Sematophyllum demissum*) P. Rhind; p. 83 (*Campylopus setifolius*) P. Rhind; p. 83 (*Nowellia curvifolia*) P. Rhind; p. 85 (*Pseudoscleropodium purum*) P. Rhind; p. 85 (*Sphagnum* spp) P. Rhind; p. 87 (*Polytrichum commune*) P. Rhind; p. 88 (*Scorpidium scorpioides*) P. Rhind; p. 88 (*Aulocomnium palustre*) P. Rhind; p. 89 (*Philonotis fontana*) P. Rhind; p. 89 (*Anthelia julacea*) P. Rhind; p. 90 (*Lunularia cruciata*) P. Rhind; p. 99 (Crimson Wax Cap) P. Rhind; p. 101 (*Galerina tibiicystis*) C. Aron; p. 101 (Sulphur Tuft) P. Rhind; p. 102 (Bog Beacon) P. Rhind; p. 103 (Honey Fungus) P. Rhind; p. 105 (*Xerocomus parasiticus*) P. Rhind; p. 106 (Beafsteak Fungus) P. Rhind; p. 106 (Many-zoned Polypore) D. Evans; p. 108 (*Cantharellus tubiformis*) P. Rhind; p. 109 (Fly Agaric) D. Evans; p. 110 (*Cortinarius paleaceus*) P. Rhind; p. 111 (Yellow Swamp Russula) P. Rhind; p. 112 (*Suillus bovinus*) P. Rhind; p. 113 (*Cordyceps capitata*) P. Rhind; p. 114 (*Peziza badia*) P. Rhind; p. 116 (Yellow Brain Fungus) P. Rhind; p. 120 (Lichens on Glyder Fach) P. Rhind; p. 122 (*Lobaria virens*) P. Rhind; p. 123 (*Pyrenula macrospora*) P. Rhind; p. 123 (*Sticta canariensis*) P. Rhind; p. 124 (*Degelia plumbea*) P. Rhind; p. 124 (*Ochrolechia tartarea*) P. Rhind; p. 125 (*Usnea florida*) P. Rhind; p. 127 (*Xanthoria parietina*) P. Rhind; p. 128 (*Ophioparma ventosa*) P. Rhind; p. 128 (*Parmelia conspersa*) P. Rhind; p. 130 (*Stereocaulon vesuvianum*) P. Rhind; p. 131 (*Cladonia coccifera*) P. Rhind; p. 131 (*Cladonia subulata*) P. Rhind; p. 132 (*Peltigera membranacea*) P. Rhind; p. 140 (Llyn Padarn) D. Evans.

INDEX

Aberdyfi 23

Aberglaslyn Pass (The) 35, 50, 53, 107

Abermaw (Barmouth) 53

Acarospora
    fuscata 126
    sinopica 132

Acer pseudoplatanus 32, 122

Achillea millefolium 28

Achnanthes 147
    lanceolata 144
    levanderi 145
    marginulata 148
    minutissima 144, 145, 148
    minutissima var. *cryptocephala* 145, 146

Actinothyrium graminis 102

Adder's-tongue 28, 52

Adelanthus decipiens 81

Adoxa moschatellina 33

Afon Cegin 138

Afon Conwy 29

Afon Eidda 29

Afon Glaslyn 29, 32, 68

Afon Llugwy 29

Afon Mawddach 123

Afon Menai (Menai Strait) 104

Afon Ogwen 143

Afon Prysor 29

Afon Rhaeadr Fawr 50

Afon Rhyddallt at Cwm y Glo 58

Afon y Llan (Coed Cochwillan) 90

Agrostis
    canina 8
    capillaris 27, 28

Alchemilla glabra 9, 28

Alder 30, 32

Alder Buckthorn 30

Alectoria nigricans 134

Aleuria
    aurantia 96
    luteonitens 97

Allantoparmelia alpicola 133-34

Allen, Tim 139

Allium ursinum 33

Alnus glutinosa 30, 32

Alpine Bistort 28

Alpine Cinquefoil 17, 18

Alpine Penny-cress 33

Alpine Saw-wort 2, 6, 17-18

Alternate Water-milfoil 21

Amanita
    citrina 104
    citrina var. alba 104
    crocea 110
    fulva 110
    muscaria 109
    nivalis 99, 117
    pantherina 104
    phalloides 104
    porphyria 105
    rubescens 104
    vaginata 110
    virosa 104

Amblyodon dealbatus 74

Amblystegium saxatile 93

Amethyst Deceiver 108

Amphidium
    lapponicum 74, 78
    mougeottii 78

Amphora
    ovalis 145
    pediculus 145

Amygdalaria pelobotryon 129, 133

Anabaena spp. 141

Anastrepta orcadensis 86

Andreaea
    alpina 71, 74
    mutabilis 74
    rothii 76
    rupestris 76

Andromeda polifolia 24

Anemone nemorosa 9

Angelica sylvestris 9

Ankistrodesmus falcatus 141

Anoectangium
    aestivum 78
    warburgii 74, 78

Anomoneis
    exilis 146, 147
    exilis var. *lanceolata* 146
    vitrea 148
Antennaria dioica 7, 16, 26
Anthelia
    julacea 74, 87, 89
    juratzkana 74
Anthoxanthum odoratum 20, 28
Aphanizomenon spp. 141
Arabis petraea 16, 61
Arctoa fulvella 74
Armeria maritima 2
Armillaria mellea 103
Artemisia campestris 36
Arthrorhaphis citrinella 127
Ascocoryne sarcoides 107
Ash 30, 32, 135
Aspicilia
    caesiocinerea 126
    calcarea 130
    contorta 130
Asplenium
    x *contrei* 53
    adiantum-nigrum 53
    obovatum ssp. *lanceolatum* 53, 55
    ruta-muraria 53
    scolopendrium 42, 51
    septentrionale 39, 40, 43, 53, 55
    trichomanes ssp. *quadrivalens* 53
    trichomanes ssp. *trichomanes* 45
    trichomanes-ramosum 45
Asterionella formosa 139, 144, 148
Asteroxylon 56
Athyrium filix-femina 43, 46, 48, 49, 53
Aulacomnium palustre 87, 88
Aulacosira subarctica 148
Autumn Hawkbit 28
Awlwort 7, 20, 21

Babington, Charles 58
Baeomyces
    placophyllus 133
    roseus 130
Bangor 73, 97, 104, 138, 139
Banks, Joseph 6

Baragwanathia 56
Barmouth (Abermaw) 53
Bartramidula wilsonii 72 93
Bartramidula wilsonii 93
Batrachospermum moniliforme 138, 143
Batterbee, Rick 147
Bazzania
    tricrenata 86
    trilobata 81
Beak-sedge
    Brown 24
    White 20, 24, 33
Beddgelert 35, 41, 73, 138
Bedstraw
    Heath 20, 27
    Marsh 24
    Northern 17, 18
Beech 32
Beech Fern 44, 46, 48, 49, 50
Beechwood Sickener 107
Beefsteak Fungus 106, 107
Belonium hystrix 102
Bennett, William 40
Benoit P.M. 73, 121
Bent
    Common 27, 28
    Velvet 8
Berkeley, Miles 97
Berwyn (The) 22, 23, 25, 147
Bethesda 104, 107, 129
Betula
    nana 59
    pubescens 30, 31
Betws-y-Coed 29, 31, 33, 51, 97, 105, 108, 114, 125, 135, 141
Bilberry 26, 31, 33, 84, 110
Birch
    Downy 30, 31
    Dwarf 59
Bird's-foot-trefoil 28
Bird's-foot-trefoil (Greater) 24
Birks, H.H. 73
Birks, H.J.B. 73
Black Alpine-sedge 19
Black Bulgar 107
Blaen-y-Cwm 65

Blaenau Dolwyddelan 7
Blaenau Ffestiniog 129
Blechnum spicant 43, 46, 48, 50, 51
Blepharostoma trichophyllum 71
Blindia acuta 78
Bluebell 33, 36
Blusher (The) 104
Blushing Bracket 112, 115
Bog Asphodel 26, 33
Bog Beacon 102, 115
Bog Myrtle 24
Bog Rosemary (Bog-rosemary) 24
Bog Stitchwort 24
Bogbean 20, 21
Bolete
    Bay 114
    Brown Birch 111
    Orange Birch 111
    Red Cracking 107
Boletus
    calopus 105, 107
    edulis 114
    erythropus 105, 107
    pulverulentus 105
Bonnet Caps 103
Botrychium lunaria 28, 45, 52
Bracken 49, 52-54
Breutelia chrysocoma 72
Brewer, Samuel 6, 19, 58, 71, 119
Brick Caps 107
Brittle Bladder-fern 45, 49
Brittle Gills 103
Broome, Charles 97
Brown Roll-rim 114
Brown, Nigel 98
Bryn Cathlwyd 7
Bryn Meurig 31, 101, 104, 106, 107
Brynrefail 116
Bryoria
    fuscescens 125
    smithii 135
    subcana 135
Bryum
    alpinum 71
    mildeanum 74, 93
    muehlenbeckii 74, 75, 87

    pseudotriquetrum 87
    weigelii 74, 87
Buckler-fern
    Broad 42, 44, 46, 49, 51
    Hay-scented 48, 51
    Narrow 51, 52
    Northern 42, 45, 48
Buellia aethalea 129
Bulgaria inquinans 107
Bur-reed
    Floating 21
    Least 21, 33
Butler, Thomas 53
Butterwort Common 24, 27
Bwlch Glas 40
Bwlch Mawr 28, 68
Bwlch Sychnant (Sychnant Pass) 51, 52

Cadair Idris 3, 6, 49, 54, 79, 122, 129, 133-34
Cae Du 125
Caerhun 105
Caernarfon 142
Caernarvonshire Fern 53
Calliergon
    cordifolium 82
    cuspidatum 87
    stramineum 87
Calluna vulgaris 31, 84
Calocera viscosa 113
Caloneis
    alpestris 145
    latiuscula 145
    obtusa 145
Caloplaca
    aurantia 130
    citrina 129, 130
    heppiana 130
Calypogeia neesiana 86
Campanula rotundifolia 10, 28
Campion
    Moss 2, 6, 16, 28
    Sea 2

Campylopus
    atrovirens 72, 76
    paradoxus 84

schwarzii 76
   setifolius 82, 83
Candelariella coralliza 126
Cantharellus tubiformis 108
Capel Curig 7, 31, 53, 58, 85, 98, 103-5
   108, 111, 138, 145
Capel Curig Woods 83, 121
Cardamine pratensis 24
Carex
   bigelowii 8, 19, 28, 61
   dioica 25, 68
   echinata 33
   hostiana 25
   magellanica 23
   pauciflora 23
   pulicaris 25
   rostrata 24, 68
   atrata 19
Carnedd Dafydd 12, 19, 42
Carnedd Llewelyn 3, 7, 10, 27, 40, 43
Carneddau (The) 7-9 23-4 26, 67, 69, 90,
   134
Carum verticillatum 29
Castell y Gwynt 8
Catapyrenium lachneum 129
Catillaria chalybeia 129, 133
Catoscopium nigritum 90
Cauliflower Fungus 113
Centaurea nigra 28
Cep 114
Cephalanthera longifolia 33
Cephalozia
   catenulata 83
   massalongi 93
Cephaloziella nicholsonii 93
Cerastium
   alpinum 16
   arcticum 16
   fontanum 16
Ceterach officinarum 53
Cetraria
   commixta 134
   hepatizon 134
   islandica 133-34
Cetrelia olivetorum 122
Ceunant Cynfal 51

Ceunant Dulyn 51
Ceunant Glanrafon (Rhyd Ddu) 50
Ceunant Llennyrch 50-51
Ceunant Mawr 41, 47, 48
Chaetophora incrassata 143
Chamaemyces fraccidus 109
Chanterelle 108
Charcoal Burner 104
Chlorella spp. 141-142
Chlorococcales 141
Chrysosplenium oppositifolium 32
Cilgwyn 25
Cinclidotus mucronatus 93
Circaea
   x *intermedia* 31
   alpina 31
   lutetiana 31-32
Cirriphyllum piliferum 82
Cirsium heterophyllum 29
Cladonia
   arbuscula 130
   ciliata 130
   crispata var. *cetrariiformis* 130
   diversa (*C.coccifera* agg.) 130
   furcata 119, 130
   portentosa 130
   pyxidata 130
   rangiferina 134
   subcervicornis 127
   subulata 130-31
Clavaria fumosa 100
Clavulina cristata 113
Clavulinopsis
   corniculata 100
   helvola 100
Clayden, Stephen 121
Clitocybe spp 103
Clogwyn Du'r Arddu 16, 40-41, 45, 60-62
   79, 127
Clogwyn y Garnedd 6, 9, 12, 40-41, 79, 129
Closterium
   acerosum 142
   didymotocum 142
   parvulum 142
   striolatum 142
Cloudberry 22, 26

Club-rush (Common) 21
Clubmoss
    Alpine 56, 58-61
    Fir 56-61
    Interrupted 56, 58-59, 61
    Lesser 56-59, 61-62
    Marsh 56, 58, 61-63
    Stag's-horn 56, 58-61
Cocconeis placentula 144
Cochlearia officinalis 10
Coed Aber Artro 51
Coed Benglog 122
Coed Bryn-Engan near Capel Curig 105
Coed Cae-Awr 121
Coed Camlyn 31, 103
Coed Coch (Abergele) 97
Coed Cochwillan 90, 104, 107
Coed Crafnant 103, 122
Coed Cymerau 49, 103
Coed Dinorwig 31
Coed Dolgarrog 30, 32, 97, 107, 109
Coed Ganllwyd 30, 51, 72, 83-84, 107, 115, 122
Coed Gerddibluog in Cwm Bychan 122
Coed Gordderu 125
Coed Gors y Gedol 33, 51
Coed Gorswen 32
Coed Gwydir 53, 79, 98, 112, 139
Coed Llechwedd 51
Coedmawr 39, 130
Coed Padarn 98, 104, 108, 110, 112, 115
Coed Tremadog 30, 32
Coed Victoria 50, 81
Coed y Gofer 33
Coed y Rhygen 50, 81, 84, 103, 112
Coed yr Allt Goch 31
Coedydd Aber 30, 47, 68, 103, 106, 109, 115, 121, 123-24
Coedydd Abergywnant 30
Coelocaulon aculaetum 130
Collema spp 129
Collema nigrescens 135
Collybia spp 103
Coltricia perennis 112
Colura calyptrifolia 82
Conocephalum conicum 71, 87

Conocybe tenera 116
Conostomum tetragonum 74, 93
Cordyceps
    capitata 113
    militaris 100
    ophioglossoides 107, 113
Coriolus versicolor 106-07
Cornicularia normoerica 133-34
Cors Arthog 24, 86
Cors Barfog 23-24
Cors Geuallt (near Capel Curig) 65, 145
Cors Goch Trawsfynydd 24
Cors Graianog 42
Cors Gyfelog (near Pant Glas) 42, 67
Cors Ty'n y Mynydd 42
Cortinarius
    alnea 115
    armillatus 110
    cinnamomeus 114
    delibutus 110
    elatior (old name) 107
    helvelloides 115
    largus 97
    paleaceus 110
    phoeniceus 97
    pholideus 97
    pseudosalor (old name) 107
    stillatitius (*C.elatior*) 107
    trivialis 115
    uliginosus 115
Corylus avellana 30
Cosmarium
    annulatum 142
    brebissonii 142
    costatum 145
    cucumis 142
    decedens 145
    elevatum 145
    etchachensis 145
    margaritiferum 142
    ochthodes 145
    parvulum 142
    reniforme 142
    tinctum 142
Cotton-grass
    Common 29

Hare's-tail 22
Slender 36
Cow-wheat (Common) 31
Craig Las (Cadair Idris) 49
Craig yr Ysfa 19
Cranberry 20
Cratoneuron commutatum 87
Crepidotus
    luteolus 115
    mollis 115
Crepis paludosa 29
Crested Coral Fungus 113
Crested Dogstail 28
Crib Goch 6, 28, 127
Crib y Ddysgyl 28
Crimson Wax Cap 99-100
Crinalium endophyticum 141
Cross-leaved Heath 20-22, 24-26
Crottle 118
Crowberry 22, 25, 59-60
Cryptogramma crispa 20, 43-45, 49
Ctenidium molluscum 84
Cuckoo Flower 24
Cwm Bochlwyd 79
Cwm Bychan near Llanbedr 72, 122
Cwm Cau (Cadair Idris) 129
Cwm Cneifion 20
Cwm Cywion 42, 48
Cwm Du'r Arddu 52
Cwm Dulyn (Llanllyfni) 52
Cwm Dyli 73
Cwm Eigiau 69
Cwm Glas Bach 19, 46, 79
Cwm Glas (Mawr) 6, 12, 14, 16, 28-29,
    41, 42, 45, 48-49, 52, 61, 79, 84, 129,
    133, 138
Cwm Glaslyn 40
Cwm Graianog 27
Cwm Idwal 9, 12, 14-16, 19-20, 24, 38,
    41, 44-45, 48-49, 61, 68, 72, 79, 85,
    87, 97, 100, 127, 129 ,133, 139, 143

Cwm Llafar 9, 24
Cwm Llefrith 41
Cwm Meillionen (Beddgelert) 133
Cwm Pen Llafar 48

Cwm Perfedd 19
Cwm y Bedol 24
Cwm y Glo 58, 63
Cwmffynnon 61
Cwmgwared (Bwlch Mawr) 68
Cwmnantcol ravine (Harlech) 82
Cyclotella 147
    comensis 145, 148
    glomerata 139, 148
    kützingiana 147
    ocellata 147
    stelligera 148
Cymatopleura solea 144
Cymbella 145
    cesati 145
    cistula 144
    cymbiformis 144
Cynosurus cristatus 28
Cystocoleus ebeneus 127
Cystoderma amianthinum 100
Cystolepiota bucknalli 109
Cystopteris fragilis 45, 49

Dactylorhiza
    incarnata 29
    maculata 29
    purpurella 29
Daedalea quercina 107
Daedaleopsis confragosa 112, 115
Daldinia concentrica 115
Davies, Hugh 119
Dead Man's Fingers 107
Death Cap 104
Deergrass 22, 26
Degelia plumbea 124-25
Deptford Pink 36
Deschampsia
    cespitosa ssp. *alpina* 1, 11, 18
    flexuosa 8, 20, 27, 31
Desmidiales 142
Destroying Angel 104
Devil's-bit Scabious 11
Dianthus armeria 36
Diatoma
    hiemale 144
    vulgare 144

Dicranodontium denudatum 81
Dicranoweisia crispula 75-76
Dicranum
    majus 81
    scottianum 72
Digitalis purpurea 33
Dillenius, John Jacob 6, 58, 71, 119, 138
Dinas Emrys 50
Dinas Mawddwy 66
Dinobryon spp. 139
Dinorwig Quarry 79
Diphasiastrum alpinum 56, 58-60
Diplophyllum
    albicans 76, 86
    taxifolium 74, 93
Diploschistes scruposus 126
Ditrichum
    plumbicola 79
    zonatum 75
Dixon, H.N. 73
Dobbs, Geoffrey 98
Docidium baculum 143
Dog Lichens 129
Dog's Mercury 32
Dolfriog Woods 121
Dolgellau 29, 68, 73, 115, 138
Dolmelyllyn 135
Draba incana 17-18
Draparnaldia glomerata 143
Drepanocladus
    revolvens 87
    vernicosus 93
Drepanolejeunea hamatifolia 82
Drepanophycus 56
Drosera
    anglica 6, 24
    intermedia 6, 24
    rotundifolia 6, 20, 23
Dryas octopetala 7, 17
Dryopteris
    aemula 48, 51
    affinis 46, 48-49, 52
    affinis ssp. *borreri* 50, 52
    carthusiana 51-52
    dilatata 42, 44, 46, 49, 51
    expansa 42, 45, 48

    filix-mas 37, 46, 49
    oreades 45, 49
    submontana 49
Duckett, J.G. 73
Dung Roundhead 116
Dutch Rush 64
Dwarf Willow 2, 8, 61, 98-99
Dyffryn Conwy 51-52, 68, 84, 97, 105, 109
Dyffryn Crafnant 7
Dyffryn Ffestiniog 29
Dyffryn Ogwen 9, 21, 44, 132
Dyfi Estuary 23

Earth Ball (Common) 105
Eifl (Yr) 48
Elaphomyces
    granulata 113
    muricatus 107
Elatine hexandra 21
Eleocharis
    palustris 21
    parvula 36
Elidir Fach 60-61, 89, 99
Elm English 32
Elm Wych 32
Empetrum nigrum 22, 25, 59-60
Encalypta alpina 73, 75, 93
Enchanter's-nightshade 31, 32
Enchanter's-nightshade Alpine 31
Enchanter's-nightshade Intermediate 31
Entoloma 97, 100
    bloxamii 100, 117
    chalybaeum 100
    conferendum 100
    serrulata 100
Ephebe lanata 132-33
Ephemerum serratum 90
Epilobium
    alsinefolium 7, 17-18
    brunnescens 35
Epipactis helleborine 32
Equisetum
    arvense 64-67
    fluviatile 21, 64-67
    hymenale 64

palustre 64-68
pratense 65-66
sylvaticum 64-69
telmateia 64-66, 68-69
variegatum 66
x litorale 64 67
Eremonotus myriocarpus 74
Erica
    cinerea 25, 84
    tetralix 20-22, 24-26
Eriophorum
    angustifolium 29
    gracile 36
    vaginatum 22
Eryngium campestre 36
Esgair-Gawr (near Dolgellau) 65
Euastrum 145
    binale 142
    cuneatum 142
    elegans 142
    montanum 145
    pectinatum 142
Eucapsis alpina 144
Eudorina spp. 141
Eunotia 145
    arcus 144
    exigua 144
    lunaris 148
    veneris 148
Euphrasia 2
    cambrica 2, 28
    rivularis 29
Eurhynchium striatum 82
Eyebright 2

Fagus sylvatica 32
Fairy Glen Woods 97, 111, 121, 135
Fallopia japonica 35
False Death Cap 104
Felin Hen 65
Fescue
    Red 28, 33
    Sheep's 8, 20, 27-28
    Viviparous 1, 8, 11, 20
Festuca
    ovina 8, 20, 27-28

    rubra 28, 33
    vivipara 1, 8, 11, 20
Ffynnon Llyffant on Carnedd Llewelyn 144
Ffynnon Oer on Snowdon 144
Field Eryngo 36
Field Wormwood 36
Filago minima 33
Filipendula ulmaria 32
Filmy-fern
    Tunbridge 41-42, 50-51
    Wilson's 41-42, 50
Fissidens
    adianthoides 87
    algarvicus 90
    cristatus 78
    polyphyllus 72
    serrulatus 93
Fistulina hepatica 106-07
Floating Water-plantain 21, 36
Fly Agaric 109
Foel Grach 7
Fomes fomentarius 112
Fontinalis squamosa 87
Fossombronia
    husnotii 90
    maritima 90
Foxglove 33
Fragilaria 145
    brevistriata 145
    capucina 144, 148
    construens 145
    construens 145
    crotonensis 139, 144, 148
    elliptica 145
    pinnata 145
    virescens 144
Frangula alnus 30
Fraxinus excelsior 30, 32, 135
Frullania tamarisci 77, 83
Frustulia rhomboides 144, 148
Funnel Caps 103
Fuscidea 129
    cyathoides 126
    kochiana 126
    lygaea 126, 129
    recensa 126

Galerina
    hypophaea 98-99
    paludosa 100
    tibiicystis 100-101
Galium
    boreale 17-18
    odoratum 33
    palustre 24
    saxatile 20, 27
Gallt y Wenallt 129
Gallt yr Ogof 48, 61, 128
Ganoderma applanatum 108
Genicularia elegans 142
Genista anglica 29
Geoglossum
    fallax 100
    viscosum 100
Gerronoma
    alpinum 99
    ericetorum 99
    hudsonianum 99, 119
Geum rivale 9
Glanllynnau (near Criccieth) 65
Gleocapsa spp. 141
Globe-flower 3, 6, 9, 29, 33
Glyder Fach 8, 120, 133
Glyder Fawr 58
Glyderau (The) 7-8, 19, 42, 56, 58, 84, 89, 99
Glyphomitrium daviesii 93
Golden-rod 9
Gomphidius
    glutinosus 97
    roseus 112
Gomphonema spp. 145
Gors Geuallt 59
Graphis
    elegans 122
    scripta 122
Greater Spearwort 24
Griffith, J.E. 42-43, 58, 72, 120
Griffith, J.Wynne 72, 119
Griffiths, B.M. 139
Grisette 110
Grimmia
    atrata 79

    elongata 75
    torquata 78
Gwydir Forest 51
Gyalecta jenensis 129
Gymnadenia conopsea 29
Gymnocarpium
    dryopteris 47-50
    robertianum 49
Gymnocolea
    acutiloba 73, 86, 93
    inflata 84
Gymnomitrion
    concinnatum 74
    coralloides 74, 93
    crenulatum 74, 76
    obtusum 74, 76
Gymnomyces xanthosporus 97
Gymnopilus junonius 113
Gyrn 131
Gyroporus cyanescens 111
Gyrosigma spp. 144

Haematomma ochroleucum 129
Hafod Garregog 31
Hafod-y-Meirch 65
Hair-grass
    Alpine Tufted 1, 11, 18
    Wavy 8, 20, 27, 31
Hairy Stereum 112
Halfway Bridge 104
Hammarbya paludosa 7, 20
Haplomitrium hookeri 70, 90
Hard-fern 43, 46, 48, 50-51
Harebell 10, 28
Harlech 51, 82
Harpalejeunea ovata 82
Hart's-tongue 42, 51
Hawker, Lilian 97
Hawkweeds 19
Hazel 30
Heath Milkwort 27
Heather
    Bell 25, 84
    Common 20-21, 25
Hedgehog Fungus 108

Helleborine
    Broad-leaved 32
    Narrow-leaved 33
Herbertus
    aduncus 71, 72, 86
    aduncus ssp. *hutchinsiae* 86
    stramineus 74, 78
Heterodermia obscurata 135
Hieracium 19
    holosericeum 19
    snowdoniense 19
Hildenbrandtia rivularis 143
Hill, M.O. 73, 78
Hirnant Valley (Rhos-y-Gwyliau) 69
Hoary Whitlowgrass 17-18
Holly 30
Holly-fern 39-40, 43, 45
Honey Fungus 103
Horsetail
    Field 64-67
    Great 64-66, 68-69
    Marsh 64- 68
    Shady 65-66
    Shore 64, 67
    Variegated 66
    Water 21, 64-67
    Wood 64-69
Huperzia selago 56-61
Hyacinthoides non-scripta 33, 36
Hyalotheca dissiliens 142
Hydnellum 108
    ferrugineum 97, 117
    scrobiculatum 117
Hydnum repandum 108
Hydrurus foetidus 145
Hygrobiella laxifolia 78
Hygrocybe 116
    calyptriformis 100, 117
    cantharellus 100
    ceracea 99
    chlorophana 99
    coccinea 99
    coccineocrenata 100
    helobia 99
    psittacina 100
    punicea 99

    reidii 100
    spadicea 100, 117
    turunda 100
Hygrohypnum
    eugyrium 87
    ochraceum 87
Hylocomium
    brevirostre 82
    splendens 80-81, 84
    umbratum 81
Hymenelia lacustris 133
Hymenogaster
    hessei 97
    tener 97
    vulgaris 97
Hymenophyllum
    tunbrigense 41-42, 50-51
    wilsonii 41-42, 50
Hyocomium armoricum 87
Hypericum undulatum 24
Hypholoma 100
    elongatum 100
    ericaeum 100
    fasciculare 100-101
    sublateritium 107
Hypnum
    andoi 83
    hamulosum 75, 78
    jutlandicum 86
    lacunosum 77
Hypocenomyce caradocensis 125
Hypogymnia physodes 125-26

Iceland Moss 134
Ilex aquifolium 30
Impatiens noli-tangere 32
Imshaugia aleurites (*Parmeliopsis*) 135
Initia vermicularis 70
Inocybe
    auricoma 97
    calaministrata 113
    godeyi 97
Inonotus radiatus 115
Isoetes
    echinospora 56, 58-59, 61, 63
    lacustris 56, 58-59, 61

Isopterygiopsis muelleranum 75, 78
Isopterygium pulchellum 78
Isothecium myosuroides 83
Ivy-leaved Bellflower 7

Japanese Knotweed 35
Jelly Babies 108
Jelly Tongue 113
Johnson, Thomas 16
Jones, D.A. 73, 120
Jubula hutchinsiae 82
Juncus
    acutiflorus 24
    effusus 24
    squarrosus 68
    triglumis 6, 19
Jungermannia borealis 74-75, 78
Juniper 1, 27
Juniperus
    communis 1, 27
    communis ssp. *nana* 27

Kershaw, Kenneth 121
Kiaeria
    blyttii 75-76
    falcata 75
Killarney Fern 37, 40-41, 43, 55
King Alfred's Burnt Cakes 115
Knapweed Common 28
Kurzia
    pauciflora 86
    trichoclados 86

Laccaria
    amethystina 108
    bicolor 98
Lachnum
    apalum 102
    diminutum 102
Lactarius 103
    aspideus 115
    blennius 107
    chrysorrheus 103
    deliciosus 113
    fuliginosus 104
    glyciosmus 110

    obscuratus 115
    pallidus 107
    quieticolor 113
    quietus 103
    repraesentaneus 98, 110
    rufus 113
    tabidus 110
    torminosus 110
    turpis 110
    uvidus 110
    variicolor 111
    vellereus 103
    vietus 110
    volemus 98, 103
Lady's-mantle 9, 28
Lady-fern 43, 46, 48-49, 53
Laetiporus sulphureus 107
Lasallia pustulata (Rock Tripe) 126-27
Lecanora
    achariana 135
    gangaleoides 129
    intricata 126
    piniperda 125
    sulphurea (*Lecidea*) 129
Leccinum 111
    holopus 98, 111
    scabrum 111
    variicolor 111
    versipelle 111
Lecidea
    lactea 129
    phaeops 133
Lecidoma demissum 134
Leighton, William Allport 120
Leiocolea heterocolpos 78
Lemanea fluviatilis 138, 143
Lemon-scented Fern 48-50
Lenzites betulina 112
Leontodon autumalis 28
Leotia lubrica 108
Lepidozia
    cupressina 81
    pearsonii 86
Lepiota spp. 109
Lepiota ignivolvata 109
Leptodontium reurvifolium 72

Leptogium
   brebissonii 135
   cyanescens 123
Leptoscyphus cuneifolius 83
Lesser Twayblade 7, 23
Leucanthemum vulgare 11
Leucobryum glaucum 80-81
Lhuyd, Edward 3, 5, 12, 38, 41, 58, 119
Liberty Cap 100
Liddle, M.J. 139
Lightfoot, John 6
Limestone Fern 48
Limosella australis 36
Listera cordata 7, 23
Lithographa tesserata 129
Littorella uniflora 6, 20-21
Llanberis 3, 20, 41, 50, 63, 80, 114, 116,
   129, 133
Llanberis Pass 20, 121
Llangower 65
Llanrwst 40
Llety Lloegr 66
Llewesig near Capel Curig 59
Lliwedd (Y) 27, 144
Lloydia serotina 2, 4, 12-13, 36
Llwyn y Coed 80
Llwynwcws (near Llanaber) 66
Llyn Bach 58
Llyn Bochlwyd (Glyder Fach) 67, 133
Llyn Bodgynydd 102, 110
Llyn Bychan 21, 58
Llyn Cau (Cadair Idris) 133
Llyn Clyd 145
Llyn Coedty (Carneddau) 69
Llyn Conwy 147
Llyn Cororion 31, 59
Llyn Cowlyd 7, 66-67
Llyn Crafnant 63, 143
Llyn Cwellyn 21, 141
Llyn Cwm Mynach 147
Llyn Du'r Arddu 29
Llyn Dulyn 147
Llyn Dwythwch 58-59
Llyn Eiddew Bach 147
Llyn Fynnon Lloer (Carneddau) 133

Llyn Gafr 79
Llyn Glas 6, 145
Llyn Goddionduon 21
Llyn Gwynant 47-48
Llyn Gyno (Berwyn) 147
Llyn Hir (Berwyn) 147
Llyn Idwal 7, 21, 28, 51, 61, 63, 67, 90,
   138, 144
Llyn Llagi 61, 147
Llyn Llewesig 145
Llyn Llydaw 58, 65, 141-42
Llyn Ogwen 7, 21, 58, 61, 97, 105, 139,
   144-45
Llyn Padarn 58, 61, 110, 115, 131, 139-
   41, 143-45, 147-48
Llyn Peris 3, 10, 139-40, 145, 147
Llyn T'yn y Mynydd 51
Llyn Tegid (Bala) 21, 31, 148
Llyn Teyrn 129, 143
Llyn Trawsfynydd 28-29, 86
Llyn y Cŵn 19, 56 ,58, 61, 89
Llyn y Gadair 79
Llyn yr Adar 61
Llynnau Gamallt 147
Llynnau Mymbyr 139
Lobaria 135
   amplissima 122
   pulmonaria 122
   scrobicularia 122
   virens 122
Lobelia dortmanna 6, 7, 21
Lotus
   corniculatus 28
   pedunculatus 24
Lungwort (Tree) 122
Lunularia cruciata 87, 90
Luronium natans 21, 36
Luzula sylvatica 9
Lycopodiella inundata 56, 58, 61-63
Lycopodium
   annotinum 56, 58-59, 61
   clavatum 56, 58-61
Lysimachia nemorum 32

Maentwrog 35, 105

Male-fern 37, 46, 49
    Mountain 45, 49
    Scaly 46, 48-49, 50, 52
Marchlyn Bach 33, 67
Marchlyn Reservoir 21
Marsh Fern 51, 55
Marsh Hawk's-beard 29
Marsupella
    adusta 75-76
    alpina 75
    emarginata 87
    sphacelata 75
    stableri 75
Mat Grass 20, 27, 68
Mawddach Estuary (The) 24, 86
Meadow Buttercup 11
Meadow-grass
    Alpine 1, 6, 18
    Glaucous 18
Meadow-rue
    Alpine 6, 10, 28, 61
    Lesser 10
Meadowsweet 32
Meconopsis cambrica 3, 9, 10
Meesia uliginosa 90
Megacollybia platyphylla 107
Melampyrum pratense 31
Melancholy Thistle 29
Melosira
    distans 147
    lirata 148
    varians 144
Mennegazzia terebrata 122
Mentha pulegium 36
Menyanthes trifoliata 20-21
Mercurialis perennis 32
Meridion circulare 144
Metzgeria temperata 83
Micrasterias
    denticulata 142
    oscitans 142
    truncata 142
Microcystus spp. 141
Microlejeunea ulicina 83
Migneint (The) 23, 26
Milium effusum 33

Milk Cap 103, 110
    Coconut-scented 110
    Fleecy 103
    Grey 110
    Oak 103
    Rufous 113
    Saffron Milk Cap 113
    Slimy Milk Cap 107
    Ugly Milk Cap 110
    Wooly Milk Cap 110
Miner's Bridge Betws-y-Coed 108
Minffordd 53, 68
Minuartia verna 18, 20
Mitrula paludosa 102, 115
Mnium
    hornum 70
    thomsonii 75, 78
Moel Eilio 100
Moel Hebog 40-41, 43, 48, 60, 84, 88
Moel Siabod 7, 29
Moel yr Ogof 41, 46, 71, 77, 79
Moel y Ci (Moelyci) 26, 103, 127
Molinia caerulea 22, 24, 26, 29, 68, 102
Monoraphidium spp. 141
Moonwort 28, 45, 52
Moschatel 33
Mougeotia 142
Mountain Avens 7, 17
Mountain Everlasting 7, 16, 26
Mountain Grisette 99, 117
Mountain Pansy 28
Mouse-ear
    Alpine 16
    Arctic 16
    Common 16
Mycena 103
    galopus 100
    haematopus 115
Mycoblastus sanguinarius 125
Mylia anomala 86
Myrica gale 24
Myriophyllum alternifolium 21
Myriosclerotinia
    curreyana 102
    sulcata 102
Myurella julacea 75, 93

Nant Bwlch-yr-Haiarn 52, 98, 113-14
Nant Ffrancon 19-20, 23, 31, 52, 54, 58-
    59, 65, 100, 105, 112, 124, 126, 128
Nantgwynant 7, 20, 27, 29-30, 50, 77-78,
    121-22
Nantgwynant (Cadair Idris) 122
Nant y Benglog 48, 61-62
Nantmor 53
Nardia
    compressa 87
    scalaris 86
Nardus stricta 20, 27, 68
Narthecium ossifragum 26, 33
Naucoria
    bohemica 98-99
    escharoides 115
    subconspersa 115
Navicula 145
    cryptocephala 144, 147
    frustulum 145
    heimansii 148
    minima 147
    pupula 144
    radiosa 144
    seminulum 145, 147
Neckera crispa 77
Nectria peziza 111
Neidium
    bisulcatum 145
    iridis 144
Neobulgaria pura 107
Nephroma laevigatum 125
Netrium
    digitus 142
    oblongum 142
Newman, Edward 38
Normandina pulchella 124
Northern Rock-cress 16, 61
Nowellia curvifolia 83
Nymphaea alba 21

Oak
    Pedunculate 121
    Sessile 30, 121
Oak Fern 47-50

Oak Maze Gill 107
Ochrolechia
    parella 129
    tartarea 118, 124-25
Odontoschisma
    denudatum 86
    sphagni 86
Oedipodium griffithianum 72, 77
Ogwen Falls 84
Oligotrichum hercynicum 86
Omphaliaster borealis 98-99
Omphalina spp. 97
Onychonema filiforme 143
Opegrapha
    gyrocarpa 127
    prosodea 135
    saxigena 127
Ophioglossum vulgatum 28, 52
Ophioparma
Ophioparma ventosa (*Haematomma*) 126,
    128
Orange Peel Fungus 96
Orchid
    Bog 7, 20
    Early Marsh 29
    Early-purple 32
    Fragrant 29
    Greater Butterfly 28
    Heath Spotted- 29
    Lesser Butterfly- 29
    Northern Marsh- 29
    Small White 3
Orchis mascula 32
Oreopteris limbosperma 48-50
Orthothecium rufescens 75, 78
Osmunda regalis 37, 42, 49, 51
Oudemansiella mucida 107
Ox-eye Daisy 11
Oxalis acetosella 31
Oxford Ragwort 35
Oxyria digyna 2, 6, 10

Pallavicinia lyellii 70, 86, 93
Pallaviciniites devonicus 70
Panaeolus rickenii 116

Pandorina spp. 141
Pannaria
    mediterranea 123
    pezizoides 129
Panther cap 104
Parmelia
    conspersa 127-28
    crinita 125
    laevigata 122
    mougeotii 127
    omphalodes 126
    quercina 121
    saxatilis 126
    taylorensis 122
Parmeliopsis hyperopta 125
Parrot Toadstool 100
Parsley Fern 20, 43-45, 49
Paton, J.A. 73
Paulschulzia spp. 141
Paxillus involutus 114
Pearson, W.H. 72
Peltigera
    apthosa 129
    canina 130
    collina 123
    lactucifolia 130
    leucophlebia 129
    membranacea 130, 132
    venosa 129, 135
Pen y Groes 25
Pen yr Ole Wen 26
Penmachno 28
Pennyroyal 36
Penrhyn Gwyn 66
Penrhyn Quarry 79
Pensychnant 112
Peronia fibula 144
Persicaria vivipara 28
Pertusaria
    corallina 129
    flavicans 129
    lactea 129
Petty Whin 29
Peziza badia 113
Phellodon 108
    confluens 97

    melaleucus 97, 117
    niger 97, 117
Philonotis
    fontana 87, 89
    seriata 75, 87
    tomentella 75
Pholiota squarrosa 115
Physcia clementei 120
Pillwort 51, 55
Pilularia globulifera 51, 55
Pine
    Corsican 112
    Scots 112
Pinguicula vulgaris 24, 27
Pinnularia spp. 144-45
Pinus
    nigra var. *maritima* 112
    sylvestris 112
Piptoporus betulinus 111
Placopsis gelida 129, 132
Plagiochila
    asplenioides 82
    atlantica 82
    corniculata 82
    porelloides 78
    punctata 83
    spinulosa 81
Plagiothecium
    curvifolium 84
    denticulatum var. *obtusifolium* 75
    platyphyllum 75
    undulatum 84
Platanthera
    bifolia 29
    chlorantha 28
Platismatia glauca 125
Pleurozium schreberi 84, 86
Pluteus xanthophaeus 115
Poa
    alpina 1, 6, 18
    glauca 18
Pohlia
    bulbifera 90
    cruda 78
    ludwigii 75
    nutans 84

Polypodium
  interjectum 46, 48, 50
  vulgare 47-48, 50
Polypody
  Common 47-48, 50
  Intermediate 46, 48, 50
Polypore
  Many-zoned 106, 107, 112
  Sulphur 107
Polyporus varius var. *nummuliformme* 115
Polystichum
  aculeatum 47, 49, 51
  lonchitis 39-40, 43, 45
  setiferum 51
Polytrichum
  alpinum 71
  commune 87
  piliferum 78-79
Pondweed
  Bog 24
  Broad-leaved 21
Pont Aberglaslyn 72
Porcelain Fungus 107
Porina
  hibernica 123
  lectissima 133
Porphyrellus pseudoscaber 105
Porpidia
  crustulata 127
  hydrophila 132
  macrocarpa 127, 129, 132
  speirea 129
  tuberculosa 126-27, 132
Porthmadog 20, 53
Potamogeton
  natans 21
  polygonifolius 24
Potentilla
  crantzii 17-18
  erecta 27, 29
Preissia quadrata 78
Priddle, J. 139
Protopteridium 37
Protosphagnum 70
Prunella vulgaris 28

Pseudephebe pubescens 133-34
Pseudephemerum nitidum 90
Pseudocraterellus sinuosus 108, 117
Pseudocyphellaria norvegica 135
Pseudohydnum gelatinosum 113
Pseudoleskeella catenulata 75
Pseudorchis albida 3
Pseudoscleropodium purum 84-85
Psilocybe semilanceata 100
Psora decipiens 129
Pteridium aquilinum 49, 52-54
Pterigynandrum filiforme 75, 93
Pterogonium gracile 77
Pulveroboletus lignicola 114, 117
Purple Moor-grass 22, 24, 26, 29, 68, 102
Purple-black Earth Tongue 97
Pyrenopsis granatina 133
Pyrenula macrospora 122, 123
Polygala serpyllifolia 27

Quercus
  petraea 30, 121
  robur 121
Quillwort 56, 58-59, 61
Quillwort Spring 56, 58-59, 61, 63

Racodium rupestre 129
Racomitrium
  affine 76
  aquaticum 76
  ellipticum 78
  ericoides 71
  fasciculare 76, 79
  lanuginosum 76, 79, 86, 134
  macounii 75
Radula voluta 82
Ralfs, John 138
Ramaria formosa 97
Ramsbottomia lamprosporoides 97
Ramsons (Wild Garlic) 33
Ranunculus
  acris 11
  lingua 24
Raspberry 26
Ratcliffe, D.A. 73
Ray, John 3, 10, 58

Razor Strop Fungus 111
Red Clover 28
Reindeer Moss 134
Rhaeadr Ewynnol (Swallow Falls) 97,
   108, 125
Rhinogau 20, 26, 84, 86
Rhizocarpon
   alpicola 134
   distinctum 129
   geographicum 126-27
   lavatum 133
   lecanorinum 129
   oederi 132
Rhizomnium punctatum 82
Rhizosolenia eriensis 147
Rhododendron 35, 117
Rhododendron ponticum 35, 117
Rhodomonas spp. 139
Rhos-y-Gwaliau 69
Rhoslefain 66
Rhyd Ddu 50
Rhynchospora
   alba 20, 24, 33
   fusca 24
Rhytidiadelphus
   loreus 81
   squarrosus 84
   subpinnatus 82, 93
Rhytidium rugosum 84
Riccardia palmata 83
Riccia
   crozalsii 73
   nigrella 93
Richards, P.W. 73
Richardson, Richard 4-5
Rigid Buckler-fern 49
Rivularia biasolettiana 138
Rock Tripe 126-27
Rose Spike Cap 112
Rose, Francis 122
Roseroot 6, 10
Rowan 30
Rowbotham, J. F. 40
Rowen 68
Roy, J. 138
Roya cambrica 142

Royal Fern 37, 42, 49, 51
Rubus
   chamaemorus 22, 26
   idaeus 26
Rumex acetosella 28
Rush
   Heath 68
   Sharp-flowered 24
   Soft 24
   Three-flowered 6, 19
Russula
   Blackening 104
   Blackish-purple 104
   Common Yellow 104, 107, 111, 113
   Fetid 104
   Geranium-scented 107
   Yellow Swamp 111
Russula 103
   atropurpurea 104
   betularum 111
   claroflava 111
   cyanoxantha 104
   cyanoxantha var. *peltereaui* 104
   emetica 104, 113
   emeticella 104
   fellea 107
   foetens 104
   integra 97
   laurocerasi 104
   lepida 107
   lutea 104
   mairei 107
   nigricans 104
   nitida 111
   ochroleuca 104, 107, 111, 113
   pumila 115, 117
   sanguinea 113
   sardonia 113
Rustyback Fern 53

Salesbury, William 2, 42
Salix herbacea 2, 8, 61, 98-99
Salwey, T 72, 120
Sanicle 33
Sanicula europaea 33
Saussurea alpina 2, 6, 17-18

Saw-wort 29
Saxifraga
    hypnoides 6, 14
    nivalis 14-15
    oppositifolia 6, 13-14, 20, 28
    rosacea 14
    stellaris 3, 6, 14-15, 24
    cespitosa 2, 11-12, 36
Saxifrage
    Alpine 14-15
    Irish 14
    Mossy 6, 14
    Opposite-leaved Golden 32
    Purple 6, 13-14, 20, 28
    Starry 3, 6, 14-15, 24
    Tufted 2, 11-12, 36
Scapania
    aequiloba 75, 78
    calcicola 75
    gymnostomophila 75
    gymnostomophila 78
    nimbosa 73, 75, 93
    ornithopodioides 71, 75
    paludosa 75
    uliginosa 75, 87
    undulata 87
Scarlet Caterpillar Fungus 100
Scarlet Hood 99
Scenedesmus spp. 142
Scenedesmus graheinsii 141
Schaereria fuscocinerea (*S. tenebrosa*) 126-27
Schistidium trichodon 75, 93
Schoenoplectus lacustris 21
Scleroderma citrinum 105
Scorpidium scorpioides 87-88
Scurvy-grass 10
Scytonema myochrous(*Conferva*) 138
Sedge
    Black Alpine- 19
    Bottle 24, 68
    Dioecious 25, 68
    Few-flowered 23
    Flea 25
    Star 33
    Stiff 8, 19, 28, 61

Tall Bog- 23
Tawny 25
Sedum
    anglicum 20
    forsterianum 20
    rosea 6, 10
Selaginella selaginoides 56-59, 61-62
Selaginellites 57
Selfheal 28
Seligeria brevifolia 78
Sematophyllum
    demissum 82, 93
    micans 82
Senecio squalidus 35
Serratula tinctoria 29
Shepherd's Cress 28
Shield-fern
    Hard 47, 49, 51
    Soft 51
Shore-weed 6, 20-21
Sickener 104
Silene
    acaulis 2, 6, 16, 28
    uniflora 2
Six-stamened Waterwort 21
Skeletonema spp. 147
Skeletonema subsalum 147
Slippery Jack 113
Small Cudweed 33
Small-leaved Lime 32
Smith, A.J.E. 73
Smith, James Edward 119
Snowdon (Yr Wyddfa) 2-3, 6, 10, 13, 25, 27-29, 40, 42-43, 58, 72, 84, 97, 100-101, 121, 129, 133-35, 138-39, 143
Snowdon Lily (Spiderwort) 2, 4, 12-13, 36
Soft Slipper Toadstool 115
Solidago virgaurea 9
Solorina
    crocata 134
    saccata 133
    spongiosa 133
Sorbus aucuparia 30
Sorrel
    Mountain 2, 6, 10

Sheep's 28
Sparassis crispa 113
Sparganium
    angustifolium 21
    natans 21, 33
Sphagnum
    auriculatum 87
    capillifolium 84, 86
    compactum 84
    contortum 87
    cuspidatum 86
    imbricatum ssp. *austinii* 86
    palustre 86
    papillosum 86
    quinquefarium 86
    recurvum 87
    subnitens 86
    tenellum 84
    teres 87
    warnstorfii 87
Spike-rush
    Common 21
    Dwarf 36
Spiked Speedwell 36
Spilonema paradoxum 133
Spirogyra 142
Spirotaenia spp. 142
Splachnum
    ampullaceum 86
    sphaericum 86
Spleenwort
    Black 53
    Common Maidenhair 53
    Forked 39-40, 43, 53, 55
    Green 45
    Lanceolate 53, 55
    Maidenhair 45
Sporastatia
    polyspora 134
    testudinea 134
Spring Sandwort 18, 20
Stag's Horn Fungus 113
Staurastrum
    margaritaceum 142
    paradoxum 142
    planktonicum 143

    punctulatum 142
Staurothele fissa 133
Stellaria alsine 24
Stereocaulon
    leucophaeopsis 132
    vesuvianum 129-30, 132-33
Stereum hirsutum 112
Sticta
    canariensis 122-23, 135
    fuliginosa 123
    limbata 119, 123
    sylvatica 123
Stigeoclonium tenue 143
Stonecrop
    English 20
    Rock 20
Strangospora pinicola 135
Stropharia semiglobata 116
Subularia aquatica 7, 20-21
Succisa pratensis 11
Suillus
    bovinus 112
    luteus 113
    variegatus 113
Sulphur Tuft 100-101
Sundew
    Great 24
    Long-leaved 6, 24
    Round-leaved 6, 20, 23
Swallow Falls (Rhaeadr Ewynnol) 97, 108, 125
Sweet Vernal Grass 20, 28
Sycamore 32, 122
Sychnant Pass (Bwlch Sychnant) 51-52
Synechococcus spp. 141
Synechocystis spp.141
Synedra rumpens 147

Tabellaria 147
    fenestrata 144
    flocculosa 139, 144-45, 146-148
    quadriseptata 148
Tal y Llyn Pass 20
Tal-y-Fan 7
Talybont Wood 66
Tan y Bwlch 53, 104

Tawny Grisette 110
Taxus baccata 1
Teesdalia nudicaulis 28
Tephrocybe palustris 100
Tephromela atra (Lecanora) 129
Tetmemorus granulatus 142
Tetracylus lacustris 144
Tetraplodon
    augustatus 75, 93
    mnioides 86
Thalictrum
    alpinum 6, 10, 28, 61
    minus 10
Thamnobryum alopecurum 87
Thamnolia
    vermicularis 120, 135
    vermicularis ssp. *subuliformis* 133
Thelidium papulare 133
Thelypteris
    palustris 51 55
    phegopteris 44, 46, 48-50
Thermutis velutina 133
Theumenidium atropurpureum 97
Thlaspi caerulescens 33
Thrift 2
Thuidium tamariscinum 81
Thymus polytrichus 28
Tilia cordata 32
Tinder Fungus 112
Tir Stent 29, 68
Toninia thiopsora (T. pulvinata) 129
Tormentil 27, 29
Tortella tortuosa 78
Tortula
    cuneifolia 90
    princeps 78
Touch-me-not Balsam 32
Tough Shanks 103
Trametes versicolor 112
Trapeliopsis
    pseudogranulosa 130
    wallrothii 129
Trefriw 97
Tregarth 26, 80, 88, 99, 102, 110
Tremadog cliffs 20
Tremella mesenterica 115

Tremolechia atrata 132
Trichocolea tomentella 82
Tricholoma psammopus 98
Trichomanes speciosum 37, 40-41, 43, 55
Trichophorum cespitosum 22, 26
Trifolium pratense 28
Tritomaria exsecta 83
Trollius europaeus 3, 6, 9, 29, 33
Truffle
    Hart's 107, 113
    Nut 97
Tryfan 19, 76
Trygfylchau 5-6, 39
Turner, Dawson 119, 138
Tweed, Ronald Duncan 138-39, 144
Twll Du (Cwm Idwal) 9-10, 12, 20
Tylothallia biformigera 129
Tyn-y-Coed near Dinas Mawddwy 66

Ulex gallii 26
Ulmus
    glabra 32
    procera 32
Ulothrix zonata 143
Umbilicaria
    cylindrica 133-34
    polyphylla 126-27
    polyrrhiza 119
    proboscidea 134
Usnea
    articulata 135
    ceratina 125
    filipendula 125
    florida 125
Vaccinium
    myrtillus 26, 31, 33, 84, 110
    oxycoccus 20
Veronica spicata 36
Verrucaria
    calciseda 130
    hochstetteri 130
    nigrescens 130
Vicia orobus 28
Viola
    lutea 28
    palustris 24, 27

riviniana 28
Violet
    Common Dog- 28
    Marsh 24, 27
Volvox spp. 141

Wahlenburgia hederacea 7
Wall-rue 53
Water Avens 9
Water Lobelia 6-7, 21
Watling, Roy 98
Watson, Walter 120
Wavy St John's-wort 24
Welsh Mudwort 36
Welsh Poppy 3, 9-10
West, William & George 138-39, 142
Western Gorse 26
White Water-lily 21
Whorled Caraway 29
Wild Thyme 28
Wild, C.J. 72
Williams, John Lloyd 41, 43
Williams, William 40
Willowherb
    Chickweed 7, 17-18
    New Zealand 35
Wilson, William 40, 42, 58, 72, 119
Wnion near Dolserau 65
Wood Anemone 9
Wood Bitter Vetch 28
Wood Millet 33
Wood-rush 9

Wood-sorrel 31
Woodhead, Norman 138-39, 144
Woodruff 33
Woodsia
    Alpine 40, 43, 45-46, 55
    Oblong 40, 43, 45-46, 55
Woodsia
    alpina 40, 43, 45-46, 55
    ilvensis 40, 43, 45-46, 55
Wyddfa Yr (Snowdon) 2-3, 6, 10, 13, 25, 27-29, 40, 42-43, 58, 72, 84, 97, 100-101, 121, 129, 133-35, 138-39, 143

Xanthidium armatum 142
Xanthoria
    candelaria 126
    parietina 126, 127
Xerocomus
    badius 114
    chrysenteron 107
    parasiticus 105
Xylaria polymorpha 107

Yarrow 28
Yellow Brain Fungus 115
Yellow Pimpernel 32
Yew 1
Ynys Môn 90, 104, 138-39
Ysgolion Duon 2, 6, 9, 12, 19, 79, 133

Zosterophyllum 56
Zygnemales 142